Software Defined Networking

Software Defined Networking

Networking DESIGN AND DEPLOYMENT

Patricia A. Morreale
Kean University, Union, New Jersey, USA

James M. Anderson
GSL Solutions, Tampa, Florida, USA

CRC Press
Taylor & Francis Group
Boca Raton London New York

CRC Press is an imprint of the
Taylor & Francis Group, an **informa** business

CRC Press
Taylor & Francis Group
6000 Broken Sound Parkway NW, Suite 300
Boca Raton, FL 33487-2742

First issued in paperback 2020

© 2015 by Taylor & Francis Group, LLC
CRC Press is an imprint of Taylor & Francis Group, an Informa business

No claim to original U.S. Government works

ISBN-13: 978-1-4822-3863-1 (hbk)
ISBN-13: 978-0-367-65890-8 (pbk)

Library of Congress Cataloging-in-Publication Data

Morreale, Patricia.
    Software defined networking : design and deployment / Patricia A. Morreale, James Anderson.
        pages cm
    Summary: "Software defined networking (SDN) provides support for geographically distributed users, without having unused resources idle. This permits dynamic scaling of network services for peak times and supports service contraction during adverse operating conditions. This book presents a thoughtful discussion of the factors which led to the emergence of SDN, why SDN is vital to the growth and deployment of cloud-based computing systems, and what the impact of the SDN revolution will be. By tracing the data plane abstractions of layers into the SDN abstraction model, readers understand the importance of the SDN revolution"-- Provided by publisher.
    Includes bibliographical references and index.
    ISBN 978-1-4822-3863-1 (hardback)
    1. Software-defined networking (Computer network technology) I. Anderson, James. II. Title.

TK5105.5833.M67 2014
004.6--dc23   2014023189

Visit the Taylor & Francis Web site at
http://www.taylorandfrancis.com

and the CRC Press Web site at
http://www.crcpress.com

Contents

v

Preface

In the preparation of this book, our objective was to provide insight into software-defined networking (SDN), suitable for both experienced network professionals and new network managers. We have worked to include the latest information while presenting SDN in context with more familiar network services and challenges.

By addressing virtualization, introducing SDN functionality, implementation, and operation, we hope to provide our readers with a unique perspective of the business case and technology motivations for considering SDN solutions. By identifying the impact of SDN on traffic management and the potential for future network service growth, we feel our readers will be provided with the required knowledge to manage current and future demand and provisioning for SDN.

In Chapter 1, the importance of virtualization is discussed, particularly the impact of virtualization on servers and networks. Chapter 2 introduces SDN, with emphasis on the network control plane. Chapters 3 through 6 discuss SDN implementation, impact on service providers, legacy networks, and network vendors. Chapter 7 provides a case study on Google's initial implementation of SDN, which was well presented at the Open Network Summit in 2012 by Urs Hölzle and Amin Vahdat (Hölzle, 2012; Vahdat, 2012) and elsewhere and (we hope) succinctly summarized here. Chapter 8 investigates OpenFlow, the hand-in-glove partner of SDN; Chapter 9 looks forward, toward more programmable networks and the languages needed to manage these environments.

This book has been prepared in an accessible, conversational style, suitable for both the advanced network engineer seeking an entry point to discussion and the novice who will use this text as a starting point for further reading and investigation. Networking is a dynamic field, and our best efforts have gone into the representation we provide here. We have tried to cite the most recent publications in this book; however, the field is moving too quickly for this to be an authoritative, comprehensive treatment. We do hope it will serve as a guidepost to additional information.

Please advise us of any errors, and we will work with our publisher to correct them.

We hope our readers find this book an excellent guide to SDN, the network revolution in our midst.

Patricia Morreale
Kean University
Union, New Jersey

James Anderson
GSL Solutions
Tampa, Florida

About the authors

Patricia Morreale, PhD, is a faculty member in the Department of Computer Science at Kean University, Union, New Jersey, where she conducts research in network management and design. Since joining Kean University, she has established the Network Laboratory, building on her prior work at Stevens Institute of Technology, and has continued her research in multimedia and mobile network performance and system design.

Dr. Morreale holds a BS from Northwestern University, an MS from the University of Missouri, and a PhD from Illinois Institute of Technology, all in computer science. She holds a patent in the design of real-time database systems and has numerous journal and conference publications. With Kornel Terplan, she coauthored *The Telecommunications Handbook* and *CRC Handbook of Modern Telecommunications* (now in a second edition).

Dr. Morreale's research has been funded by the National Science Foundation (NSF), US Navy, US Air Force, AT&T, and others. She is a senior member of both the Association for Computing Machinery (ACM) and the Institute of Electrical and Electronics Engineers (IEEE) and a member of Sigma Xi. She serves as a councilor (elected) for the Mathematics and Computer Science Division of the Council on Undergraduate Research (CUR). She has served as a guest editor for *IEEE Communications* magazine, vice chair for *INFOCOM*, and on technical program committees. She is cochair of the National Center for Women and Information Technology's (NCWIT) Academic Alliance.

She has lectured internationally on network design and telecommunications service delivery. Prior to joining academia, she was in industry, working on network management and performance. She has been a consultant on a number of government and industrial projects.

Jim Anderson, PhD, is the vice president of product management at GSL Solutions. Dr. Anderson previously worked for the Boeing Aircraft company, Siemens, Alcatel, Verizon, and a number of start-ups. Dr. Anderson's research interests include effective network management techniques for very large networks, software as a service (SaaS) products, and software-defined networking/networks function virtualization (SDN/NFV).

Dr. Anderson holds a BS from Washington University in St. Louis, an MS from Washington University in St. Louis, a PhD from Florida Atlantic University, and an MBA from the University of Texas at Dallas. Dr. Anderson's BS, MS, and PhD degrees are all in computer science, and his MBA focused on marketing. He has authored over 30 books and has numerous conference publications. Dr. Anderson was a contributing author to *The Telecommunications Handbook*.

Dr. Anderson is a member of the ACM and a Senior Member of the IEEE. Dr. Anderson has served as the chairman of the IEEE's Florida West Coast Section, chairman of the IEEE's Florida Council, as well as the membership development chairman and public information officer for the IEEE's Region 3. He received the IEEE's Tampa Bay Engineer of the Year Award in 2012 and the IEEE Region 3 Outstanding Service Award in 2013.

Dr. Anderson is the president of Blue Elephant Consulting, a business communications consulting firm. In this capacity, he has spoken at many different events on topics ranging from negotiating, public speaking, product management, managing teams, and managing information technology departments. He is currently a contributing author to *CEO India*.

List of abbreviations

ACL	access control list
ALTO	application-layer traffic optimization (protocol)
AP	application program
API	application program interface
ASIC	application-specific integrated circuit
BGP-LS	border gateway protocol–link state
BGP-TE	border gateway protocol–traffic engineering
CAPEX	capital expenditure
CIO	chief information officer
CLI	command line interface
CO	central office
Codec	coder/decoder
COTS	commercial off-the-shelf software
CPU	central processing unit
CSMA	carrier sense multiple access
DB	database
DBMS	database management system
DHCP	dynamic host configuration protocol
DPI	deep packet inspection
DTLS	datagram transport layer secutity
EBGP	edge border gateway protocol
EIGRP	enhanced interior gateway routing protocol
EMS	element management system
FEC	forward error correction
FTP	file transfer protocol
GRE/UDP	generic routing encapsulation/user datagram protocol
HTTP	hypertext transfer protocol
iBGP	interior border gateway protocol
IP	internet protocol
IP-in-GRE	IP-in-generic routing encapsulation
ISDN	integrated services digital network
ISIS	intermediate system-to-intermediate system (protocol)
IT	information technology
L2/L3	level 2/level 3
LLDP	link layer discovery protocol
LVM	logical volume manager

MPLS	multiprotocol label switching (protocol)
MPLS-TE	multiprotocol label switching–traffic engineering
NFV	network function virtualization
NMS	network management system
NOC	network operations center
NOS	network operating system
NV-GRE	network virtualization using generic routing encapsulation
OFA	OpenFlow agent
OFC	OpenFlow controller
OS	operating system
OSPF	open shortest path first
OVSDB	Open vSwitch database management (protocol)
PCE	path computational element
RAID	redundant array of independent disks
RAM	random access memory
RSVP	resource reservation protocol
SANE	secure architecture for the networked enterprise
SCSI	small computer system interface
SDK	software development kit
SDN	software-defined networking
SLA	service-level agreement
SNMP	simple network management protocol
SWAN	software-driven WAN
TCP	transmission control protocol
TLS	transport-layer security
UDP	user datagram protocol
VM	virtual machine
VPN	virtual private network
WAN	wide area network
XML	extensible markup language

chapter one

Virtualization

1.1 *Introduction*

Virtualization has been an integral part of computing since distributed computing began. Users were initially provided with virtual resources, which were effectively logical subdivisions of a single shared device or devices. The concept of virtualization provided individual, dedicated resources from a larger common pool of resources and provided users with the desired customization and control. The field of virtualization expanded in part because of the limitations of shared resources. Although virtualization has been used since distributed computing started, the integration of virtualization and networking, the heart of software-defined networking (SDN), has been driven and enabled by reductions in hardware cost, advances in software, and limitations in current network configurations.

Servers and hardware costs have declined over time. When a company's information technology (IT) team created a new application, additional processing needs were handled by purchasing more servers and installing the servers in the corporate data center. Once the new hardware was assembled and available for use, often called "racked and stacked," the hardware was plugged in, with the new application software installed on the hardware, and the additional software functionality was off and running. The steep decline in the cost of server hardware beginning in the late 1990s and continuing into the first half of the twenty-first century reduced the cost of adding additional servers to corporate IT infrastructure.

This unparalleled, unchecked hardware growth began to slow once data center costs exponentially increased, as overall data center costs also include operating personnel, software licenses, maintenance costs, as well as utility and building costs. Chief information officers (CIOs) and other senior managers concluded, after reviewing data center costs, that although hardware costs had fallen, their total costs for operating a data center had risen and would continue to increase.

Corporate data centers continued expanding without constraints and were still requiring more space. All companies with IT interests were going to have to make sizable investments in building new data centers, which would be an ongoing, increasing cost. However, few of the servers that were in the existing data centers were fully utilized. In fact, many of the servers were running at less than 10% utilization. With the realization that this was an enormous waste of computing resources, it was clear that a new approach was needed to provide resources without excessive capitalization costs and with a sustainable, scalable growth plan. The answer was virtualization.

1.2   Virtual memory

In the mid-twentieth century, when mainframes first appeared in corporate processing centers, a problem quickly appeared. The programs that were being written for the mainframe computers were too large. With the software instructions needed for execution, program variables to be stored, and extra control information that the mainframe's operating system (OS) used, available memory was quickly exceeded. The IBM 650 computer (circa 1953–1969) had a memory size of only 2,000 words (a word was 36 bits) (Viking Waters, n.d.).

With the arrival of shared computing resources that supported multiple simultaneous users, memory constraints became more visible. Unlike today's memory expansion approach, for which additional memory is available and affordable, memory was both expensive and limited at the birth of mainframes because early computers could not support memory expansion. This was a significant problem for computer designers.

This problem was solved by "virtualizing" the computer's memory. The virtual memory technology used permitted a computer with a limited amount of physical memory to use the local hard disk storage system along with physical memory to make it appear that the computer had more memory available for use by application software.

Initially, virtual memory referred to the concept of using a computer's hard disk to extend a computer's physical memory. The idea was that programs running on the computer would distinguish between whether the memory was "real" memory (i.e., random access memory or RAM) or disk based. The OS and hardware would determine the actual memory location. This solved the problem of trying to run programs that were larger than the computer's memory.

Later, virtual memory was used as a means of memory protection. Every program uses a range of addresses called the address space. The use of virtual memory for memory protection solved the problem of allowing multiple users to use the same computer at the same time. The use of virtual memory prevents programs from interfering with each other. If a user's process tries to access an address that is not part of available address space, then an error occurs. The OS assumes control. The process is usually killed or terminated as a safeguard.

A computer that has been programmed to make use of virtual memory has the additional task of managing its memory. The computer will inspect RAM to determine which programs or data that have been loaded into actual physical memory space have not been used recently. Once such areas have been identified, the central processing unit (CPU) will then copy the programs from RAM to the computer's hard drive. The RAM that this information had been using is now available for use by other applications.

The copying of RAM contents to the hard drive happens both automatically and quickly. The result is that the end user and other applications are not aware that this is taking place. As a result, the entire contents of the computer's RAM storage will appear to be available for each application. The lower cost of disk space over computer memory provides an additional economic incentive for this approach.

However, there are differences between the computer's hard drive and RAM. The biggest difference is that the read/write speed of a hard drive is much slower than the read/write speed of RAM. In addition, hard drives have not been designed to support accessing small pieces of information. If there is too much interaction between the CPU and the hard drive to support the virtual memory function, the end user and associated applications will observe a slowdown in performance. The mechanics of virtual memory system implementation are very important.

1.2.1 Virtual memory operation

Virtual memory starts with a program that is too large to completely fit into the computer's physical memory. This program can be a stand-alone application, a user's online session with system configuration settings, or multiple user applications. The first thing that the computer's OS must do is divide a single large application up into equal-size chunks of the original program (called "pages") that are generally between 4 K and 64 K in size.

The computer's physical memory (RAM) is then divided into equal-sized pages. The memory addressed by a process is also divided into logical pages of the same size. When a process references a memory address, the memory manager fetches from disk the page that includes the referenced address and places it in a vacant physical page in the RAM. This is shown in Figure 1.1. Pages that are not currently used are stored on the hard disk. Subsequent references within that logical page are routed to the physical page. When the process references an address from another logical page, that new page is also fetched into a vacant physical page and becomes the target of subsequent similar references.

If the computer does not have a free physical page when it is time to swap in a virtual page, the memory manager can swap out a logical page into the hard disk's swap area and copies (swaps in) the requested logical page into the now-vacant physical page. The page that was swapped out may belong to a different process. There are many strategies for choosing which page is to be swapped out. If a page is swapped out and is later

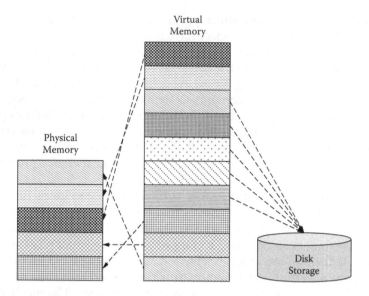

Figure 1.1 Virtual memory to physical memory mapping.

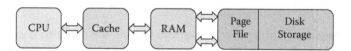

Figure 1.2 Memory management mapping.

referenced again, the page will be swapped back in from the swap area at the expense of another page that is currently in the computer's RAM.

This mapping allows the computer's main processor to view all of the memory that it is addressing as part of one contiguous address space. For each program that is running on the computer to execute correctly, it is the responsibility of the OS to manage the virtual address spaces and to assign real memory locations to virtual memory locations. The OS has the ability to create a virtual memory space that is much larger than the amount of physical memory installed on the computer. This management of the computer's virtual memory space is often performed by computer hardware called the memory management unit (MMU). This hardware is responsible for automatically translating the virtual addresses that are in use by the executing applications to physical addresses.

The area of the hard disk that stores the RAM image is called a page file. It holds pages of RAM on the hard disk, and the OS moves data back and forth between the page file and RAM. This process is shown in Figure 1.2.

The virtual memory system enables each process running on the computer to act as if it has the computer's entire memory space to itself because the addresses that it uses to reference memory are translated by the virtual memory mechanism into different addresses in physical memory. This allows different processes to use the same memory addresses on the same computer. The memory manager will translate references to the same memory address by two different processes into different physical addresses.

A process running on the computer may use an address space larger than the available physical memory. Each reference to an address from within the program will be translated into an existing physical address. The bound on the amount of memory that a process may actually address is the size of the swap area, which may be smaller than the addressable space.

An example of this would be a process that has an address space of 4 GB yet actually only uses 2 GB, and this can run on a machine with a page file size of 2 GB. The size of the virtual memory on a system must be smaller than the sum of the computer's physical RAM and the swap area. The reason for this limitation is that pages that are swapped in are not erased from the page file swap area, so they take up two pages in the total available memory space.

The Windows OS uses a swap area that is 1.5 times the size of the computer's RAM. This means that for a computer that has 4 GB of RAM, the swap area will be 6 GB, and the total size of the virtual memory will be 10 GB.

1.2.2   Virtual and physical memory mapping

The key to a successful virtual memory system is to create a process by which the computer can use virtual addresses to execute a program while the OS handles the details of mapping between the application's virtual address space and the computer's physical address space. Every program in a computer's memory, along with its data, is called a *process*. Within the computer, each process is given its own address space. An address space is a sequence of valid memory addresses that can be used by the process to store both code and variables. An address space is not fixed. While a process is executing, it can request additional memory from the OS. Once the process is done with the additional memory, the memory can then be given back to the OS.

As a process executes, it generates addresses in one of three different ways: a load instruction, a store instruction, or a fetch instruction for the next instruction to be executed. These addresses generated by the process are considered to be "virtual addresses." For actual data or instructions to be fetched or stored, this virtual address will then have to be mapped to a real physical address.

What this means is that the computer that is using virtual memory will be performing the virtual-to-physical memory mapping function all the time. To make this process as efficient as possible, it is generally handled by special-purpose hardware.

There is an added benefit to using virtual memory when more than one process is executing on a computer. The virtual memory system provides a process with memory protection via address translation.

Figure 1.3 shows how virtual memory is mapped to physical memory. It is a six-step process. The steps are as follows:

Step 1: Check an internal virtual memory address lookup table for this process. Determine if the reference was a valid request or an attempt at an invalid memory access.

Step 2: If the virtual memory address reference was invalid, terminate the process. If it was valid but the page has not yet been brought in, page in the requested page.

Step 3: Find a free frame in the computer's RAM.

Step 4: Schedule a disk operation to read the desired page into the newly allocated RAM frame.

Step 5: When the disk read is complete, modify the internal table that is kept with the process and the page table to indicate that the page is now in memory.

Step 6: Restart the process instruction that was interrupted by the address trap. The process can now access the page as though it had always been in memory.

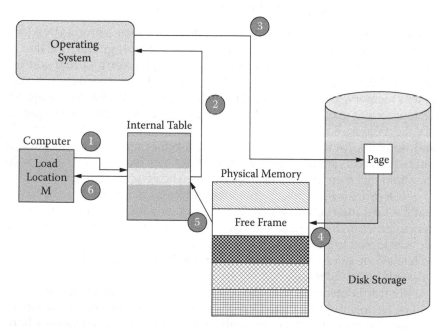

Figure 1.3 Virtual memory mapped physical memory.

The extra effort that is required to implement a virtual memory system is generally acceptable because of the advantages that the virtual memory system delivers. A virtual memory system will provide applications with increased security because of memory isolation, will free applications from having to manage a shared memory space, and will allow applications to use more memory than might be physically available using the technique of virtual memory paging.

1.3   Server virtualization

With the success found with virtualizing computer memory, computer scientists were ready to tackle their next computer challenge: how to allow multiple people to use a computer at the same time. Original computers were expensive and, with limited functionality, could only do one thing at a time. Users would queue up and wait to have their program loaded, run, and the results printed. All of this happened in a sequential fashion. Clearly, as computers became increasingly vital to the way that businesses and governments operated, the inefficiencies of sequential processing had to be resolved.

The solution to this problem was again novel: The entire computer was virtualized. The initial foray into computer virtualization took the form of the creation of a "time-sharing" mainframe system and culminated in the development of the CP-40 OS. Each user was provided with a virtual machine, which enabled multiple users to access the same mainframe computer simultaneously (Figure 1.4). A software hypervisor was created to manage memory sharing in the mainframe. Every user was fooled into believing that they were the only user on the mainframe computer at any given time. Meanwhile the mainframe raced around frantically doing small jobs for each user in a sequential fashion.

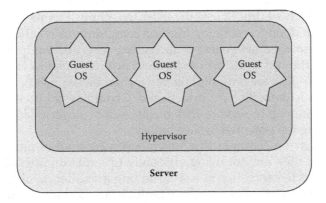

Figure 1.4 Server virtualization.

1.3.1 Importance of virtualizing servers

With the arrival of personal computers, the need to use server virtualization techniques faded away. These new computers supported only a single user, and there was no longer a need to support the overhead that server virtualization required. The single user made use of all of the personal computer's resources, and all needs were met.

As time moved on, personal computers become more powerful, and they started to be used by themselves, without human involvement. These enterprise-class servers were placed in racks, which were then placed in data centers. Soon, hundreds of servers were used to run a wide variety of business applications every day.

Many of today's servers exist in data centers. These data centers are responsible for making sure that the servers have power 7 days a week, 24 hours a day, 365 days a year (7 x 24 x 365). Making sure that all of the servers in the data center always have power is so important that most data centers create two identical but separate power systems, one the primary system and the other the secondary, or backup system. This can lead to a great deal of waste.

At the request of the *New York Times*, the consulting firm McKinsey & Company analyzed energy use by data centers and found that, on average, they were using only 6–12% of the available electricity powering their servers to perform computations (Glanz, 2012). This operating environment was a result of how the systems evolved. Initially, in the early 1990s, software systems crashed if they were asked to do too many things or even if they were turned on and off. In response, computer operators seldom ran more than one application on each server and kept the machines on around the clock, no matter how sporadically that application might be called on. Server and software operation was the goal, not energy conservation, as a crash or a slowdown could be highly disruptive to ongoing business operations. Every computer is kept up and running no matter how much (or how little) it is being used. In technical terms, the portion of a computer's brainpower being used on computations is called *utilization*. Despite heightened awareness of increasing costs and equipment underutilization, the server utilization figures have remained low: The current findings of 6–12% are only slightly better than those in 2008.

Server virtualization introduces added complexity into the operation and management of servers that businesses or government are using, while having the potential to save money.

Firms have discovered that they are paying a lot of money to keep servers that they are not using efficiently up and running all the time. Shutting selected servers down and running a smaller number of servers that have been virtualized to allow multiple applications to run on them at the same time makes good business sense.

Cost savings and reducing the number of servers that a business needs to run its business are good reasons to implement server virtualization. Additional benefits to implementing virtual servers include better ways to integrate, provision, deploy, and manage information technology (IT) systems—at a faster pace—without further straining already-tight budgets. Greater optimization and efficiency are needed in how software and solutions that power data centers are deployed and managed. Having the ability to create a new virtual machine with the press of a button can result in significant savings in both time and expense. This task is so simple to do, in some cases the creation of new virtual servers can be turned over to the actual end users.

To both streamline and optimize the way that a company's IT infrastructure operates, servers need to be virtualized and then handed out on an as-needed basis. Virtualization is not a goal by itself. A much better way to think about virtualization is to view it as a means to the strategic goal of enabling services-based IT in the enterprise.

1.3.2   Hypervisor role in server virtualization

Figure 1.5 shows the components of a nonvirtualized server, without a hypervisor. In this model, vendor-provided hardware is used for the entire server system. An OS is added that interacts with the hardware via hardware drivers, customized to the specific vendor hardware that is being used. On top of the OS, a collection of applications and services can be run independent of the underlying vendor hardware.

In this model, only one OS can be used on a hardware platform at a given point in time. In addition, the functionality that the OS is able to provide to the upper-layer applications and services is limited by the

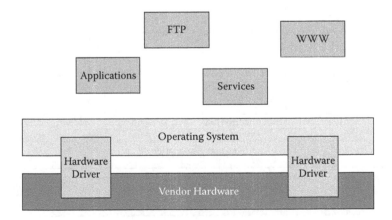

Figure 1.5 Traditional model of a nonvirtualized server.

functionality that is provided to the OS by the hardware drivers. Note that changing an OS or the underlying hardware in this configuration is difficult to do. A side effect of using this model is that a company often ends up with a collection of underutilized servers ("server sprawl") because most software vendors state that their applications should run on stand-alone servers to avoid software conflict issues with other applications.

In Figure 1.6, the architecture of a virtualized server is shown. This type of virtualization is known as "platform virtualization." A new layer of software has been added in this figure: a virtualization layer (also known as a *hypervisor*) that sits between the vendor hardware and the various OSs that will be running on this virtualized server. A large number of vendors currently offer virtualization-layer software (VMware, Microsoft, Citrix, Oracle, etc.) along with popular open-source solutions (i.e., Xen, Hyper-V, KVM, etc.).

The virtualization layer will have the needed drivers embedded in it that will control the underlying hardware. The virtualization layer then presents generic server hardware to the OS: generic graphics cards, generic network interfaces, generic storage systems, and so on.

On a virtualized server, there will be several differences from the nonvirtualized server. The first of these changes is that multiple OSs can be run on the same physical hardware at the same time. Each one of these OSs will use a piece of the available total computing resources. Each of the virtual environments (virtual machines) is both encapsulated and isolated.

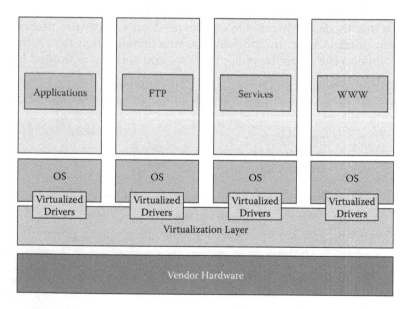

Figure 1.6 Architecture of a virtualized server.

It cannot have an impact on any of the other virtual machines while it is running. No virtual machine will be allowed to monopolize the physical server or any of its resources.

The OSs can no longer "see" the actual hardware on which they are running—the virtualization layer makes all of the hardware appear to be generic hardware. This means that each of the virtual machines is now portable. Because each of the virtual machines is defined by a set of files, each can be easily moved from one physical server to another as long as the new machine is running the same virtualization layer.

When virtual machines have been backed up, they can be restored to any hardware. This means that in the event of the hardware failure of a server in a data center, a new server from a different vendor can be installed to replace the failed server. Once the virtualization layer of software is installed on the new vendor hardware, the backups of the virtual machines that were running on the old vendor's hardware can now be restored, and they will continue to operate with no changes—the underlying hardware no longer matters.

The ability to restart a virtual machine is a fundamental feature of most virtualization layers. In the event of the failure of a virtual machine, the virtualization layer can automatically restart the virtual machine. If the entire server fails, the virtualization layer has been networked to other virtualization layers in the data center and shared storage is being used, so the virtual machines can be restarted on other servers.

Virtualization layers are able to measure the utilization of the servers on which they are running. This allows additional virtual machines to be continuously added to the server until a predefined level of utilization is reached. It is recommended that a buffer be left in this utilization level (say 60%) to allow for spikes in individual virtual machine processing loads. Note that this is much better than the estimated 4–7% utilization of most nonvirtualized servers that are found in today's data centers (US Environmental Protection Agency, 2007). In this way, companies can ensure that each of the servers in their data center is fully utilized.

Implementing virtualized servers in a data center allows servers to be consolidated. This occurs when the applications that used to run on multiple servers are now combined and start to run on a single server. Server consolidation ratios of 7x (seven physical servers replaced by one physical server running a virtualization layer) or 10x are not uncommon.

At the heart of any server virtualization is the virtualization layer of software, called the *hypervisor*. A hypervisor is a piece of software that creates the layer of virtualization that makes virtualization on a server possible. The hypervisor contains the virtual machine manager (VMM), which has the responsibility of managing each of the active virtual machines currently running on the server.

There are two different types of hypervisors. Both types are shown in Figure 1.7. The first, Type 1 ("bare metal"), is loaded directly onto the server hardware. Examples include Microsoft's Hyper-V, VMware's ESX/ESXi, and open-source Xen. This software interacts directly with the server hardware without any layer of software between it and the server hardware. Type 1 hypervisors provide true partition isolation, high reliability for virtual machines, and a high degree of security.

The other type of hypervisor, Type 2 ("OS hosted"), is loaded into the OS, which is already running on the server hardware. Examples include VMWare Workstation, Microsoft Virtual Server, and VMWare Fusion. This type of hypervisor then interacts with the underlying OS, which then interacts with the server hardware.

Type 2 hypervisors may initially appear to be easier to install than Type 1 hypervisors. Simply put, they are just an application that is run on a standard server—no special software installation is required. However, Type 2 hypervisors do not perform as well as Type 1 hypervisors simply because the OS is in between the Type 2 hypervisor and the physical hardware. Type 2 hypervisors provide low-cost virtualization, do not require the use of any additional drivers, and are easy both to install and to use.

This means that there is additional software overhead when a Type 2 hypervisor is used, which results in the restriction that it is not possible to fit as many virtual machines onto a server that is using a Type 2 hypervisor. Ultimately, the consolidation ratio that can be achieved with a Type 2 hypervisor is smaller than that which can be achieved with a Type 1 hypervisor.

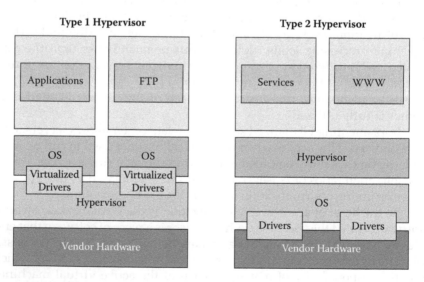

Figure 1.7 Types of hypervisors.

Type 1 hypervisors are generally used in data centers. They are used to maximize the number of physical severs that can be consolidated onto a single virtual server. Type 2 hypervisors are more often used on laptops and desktops when there is a need to run multiple OSs or create two separate development environments for testing or development purposes.

The hypervisor views the server as providing four main types of resources. These resources are the CPU, memory (RAM), storage (hard disk), and a network connection. The hypervisor then uses these four types of resources to provide the various virtual machines with the functionality that they need. This includes infrastructure services such as memory management, storage management, and network data exchange management.

Server consolidation will result in reduced costs for the company that is using the servers. These costs will include reductions in energy costs, software license costs, data center costs, and maintenance contracts. At the same time, disaster recovery will be simplified and improved, as will server manageability.

All hypervisors use a "control domain" to manage the hypervisor and the virtual machines that are running on it. This is a special-purpose virtual machine that has the ability to manage the hypervisor. In some hypervisors, such as Xen, this is also where the virtual drivers execute and device models are stored. The management console that the system administrator uses to configure the virtual server is connected to the control domain (Xen Project, n.d.).

The hypervisor management system is the key to accessing the functionality that the virtualization system provides. The basic functions of creating, halting, and terminating virtual machines are provided. In addition, the ability to create live snapshots and checkpoints and migrate virtual machines to other servers is provided. While a virtual machine is running, it can be migrated. This migration can be between host servers or pools of host servers, including those that do not have shared storage.

A hypervisor can also be regarded as a special type of application that serves to abstract the underlying physical server hardware from the guest OSs that are executing in each of the virtual machines that are running on the virtual server. Figure 1.8 shows the typical resources of a hypervisor. A hypervisor allows each application running on the server to see the resources provided by its virtual machine instead of the actual physical server resources.

One of the fundamental tasks of a hypervisor is to permit a guest OS to boot on the virtual server. To permit this to happen, the hypervisor needs to provide the OS kernel image (Windows, Linux, Solaris, etc.) with a configuration file that tells it what Internet Protocol (IP) addresses can be used, how much RAM is available, and so on. In addition, access to a hard disk for file storage and a network card for external communication also needs to be provided. The virtual disk and virtual network interface will be mapped

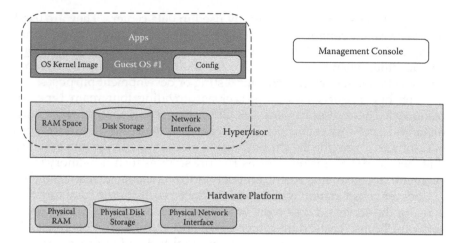

Figure 1.8 Typical hypervisor resources.

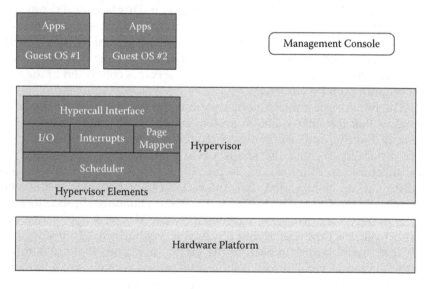

Figure 1.9 Detailed view of a hypervisor.

by the hypervisor into the server's physical disk and network interface. The one other component that is required is a management console that permits both the hypervisor and the virtual machines to be controlled.

The architecture of a hypervisor is designed to allow guest OSs to be run concurrently with the hypervisor. Figure 1.9 shows the hypervisor elements that are required. In typical applications, calls will be made to the OS to perform functions on the server's physical resources (RAM, hard disk, network card, etc.). The hypervisor provides the hypercall layer to allow guest OSs to make requests of the server's physical resources. Likewise, when an application attempts to use I/O (input/output) devices attached to the server, the hypervisor will virtualize the request and map it to the physical device.

Interrupts pose a special challenge for the hypervisor. Interrupts for virtual devices have to be routed to the guest OS. Also, during the execution of an application in a virtual machine, traps or exceptions can occur. When this happens, the impact needs only to affect the virtual machine where the trap or exception occurred and not the hypervisor or any other virtual machines.

Within the hypervisor, the page mapper component is responsible for ensuring that the server's instruction-fetching hardware always points to the correct page for the virtual machine (or the hypervisor) that happens to be executing at that time. Managing who is currently executing is the responsibility of the hypervisor's scheduler component, which is responsible for transferring execution control between the hypervisor and the various virtual machines that are executing.

Resource pools can be created that contain both flexible storage and networking resources. Within the virtual environment, events can be tracked and progress notifications can be sent. Upgrades to the virtual environment can be made and patches can be applied without having to take the virtual environment down. Real-time performance monitoring and alerting capabilities are provided.

1.3.3   Types of virtualization

There are four different ways to virtualize a server (FindtheBest.com, n.d.). Each of these approaches uses a different configuration of the three virtualization components: applications, OSs, and hypervisors.

- **Full virtualization**: When full virtualization is used, the hypervisor is responsible for completely simulating the underlying vendor hardware. This permits unmodified copies of OSs (e.g., Windows, Linux, etc.) to execute on the virtualized server within their own virtual machines.
- **Hardware-assisted full virtualization**: As virtualization has become both more popular and more critical to the effective operation of a data center, CPU manufacturers have responded by adding instructions to their products that support virtualization. When these virtualization-enabled CPUs are used to power a server, the hypervisor can leverage their features to allow guest OSs to operate in complete isolation. One feature of these CPUs is the introduction of the "ring" concept, which refers to

levels of security privileges that are permitted in the code that is currently executing. Applications operate at a ring 3 level, rings 1 and 2 are used to execute device drivers, and ring 0 is used to execute the hypervisor. AMD and Intel have also created a ring −1 that permits the hypervisor to run computations directly instead of going through the OS. This results in an increase in the efficiency of the processing.

- **Paravirtualization:** In a virtual server that is using paravirtualization, the guest OSs have each been modified to inform them that they are operating in a virtual environment. Paravirtualization permits the relocating execution of critical tasks from the virtual domain to the host domain. As a result of this, guest OSs will spend less time performing operations that are more difficult in a virtual environment compared to a nonvirtualized environment.

- **OS virtualization:** When OS virtualization is used on a server, a hypervisor is not used at all. Instead, the virtualization capability is built into the host OS. The host OS performs all the functions of a fully virtualized hypervisor. The biggest limitation of this approach is that all the virtual machines must run the same OS. Each virtual machine remains independent from all the others, but it is not possible to mix and match OSs among them.

1.3.4   Server virtualization in operation

The original problem that businesses encountered with the servers that they were using to run their enterprise applications was that too many servers were needed to do the job. When enterprise applications were created, they were designed to run on dedicated, purpose-built servers for maximum performance and stability. This was a good idea, but it led to server sprawl, where an expensive data center could quickly fill with servers that were only partially used. This is the problem that server virtualization was created to solve.

The invention of server virtualization technology now allows enterprises to consolidate both applications and OSs (jointly referred to as "workload" when talking about virtualization) that are running on multiple unrelated servers. Now, these can all be run on a single server that is executing multiple virtual environments.

The use of virtualization allows an enterprise to run multiple applications on one single physical server. Each application executes in a separate virtual environment. This allows each application to believe that it is running on a purpose-built server. Allowing each application to continue to execute on top of its specific OS eliminates a potential source of interoperability issues that could occur between applications that were required to share an OS.

1.4 Storage virtualization

Just as virtualization has had a dramatic impact on the world of servers, the world of storage needed a similar type of revolution. The amount of data generated on a daily basis that has to be stored has grown every year. EMC and the International Data Corporation together estimated that more than 1.8 trillion GB of digital information were created globally in 2011 (Glanz, 2012). Just generating and storing this data are not enough. The stored data has to be made available to servers so that software applications can process it.

1.4.1 Computer storage operation

The revolution in computer storage began in the era of mainframes. System designers had a problem: The magnetic disks that the mainframes were using were not reliable, and failures were common. They needed to find a solution to this problem. Their solution was to take large numbers of magnetic disks and pool them. This pool was then arranged to provide fault tolerance and protect against individual disk failures. This system became known as RAID (Redundant Array of Independent Disks). Applications could use a common pool of cache memory to work with a logical image of a data block rather than having to work with the actual data block as it was spun around on a disk platter. The result of this design was that it improved performance by masking the seek and rotation delays of a mechanical disk. An added benefit was that it allowed mainframes to use lower-cost disks.

At the same time, the early 1980s, desktop computers appeared. Computer storage was simple to implement for these systems. A hard drive would be added to the computer system, a cable would be run between the hard drive and the rest of the computer, and the stored information could be accessed by whichever program happened to be running. This type of solution was called direct attached storage (DAS).

The need for storage of the information that is generated daily keeps growing, and this has been true since early desktop computers. Very quickly, users started to exceed the storage limitations of their DAS solutions. The next step was for standard protocols (the language used to communicate between devices) to emerge. Examples of this included the Small Computer System Interface (SCSI). One of the strengths of this standardized way of connecting a computer to a storage system is that it made it easy to connect different vendor storage solutions to the systems.

Where the data was actually stored also continued to change. The use of SCSIs extended connectivity beyond traditional magnetic storage to devices such as CD-ROMs, tape drives and autoloaders, and JBOD (just a bunch of disks). Storage solutions continued to evolve, and soon fault-tolerant designs were available that provided greater reliability. However, the connectivity of these storage systems was still confined to single servers or workstations, limiting the utilization of the media. If a specific data set was needed, a user needed to use a specific computer to access the data set.

1.4.2   Network-attached storage

The limitation of only being able to access a data set from a single computer was a major problem. The world of computer networking was exploding at this time: Expensive proprietary solutions were giving way to a fast, simple solution called Ethernet. The ability to network multiple servers together let companies such as Novell and Sun Microsystems start to wonder if the new high-speed networking could be used to solve the limitations of the current storage systems.

Their research into this problem resulted in the creation of network-attached storage (NAS). NAS allowed a computer to treat the storage that it was using as though it was physically attached to the server even if it was in a different room or perhaps a different city. This functionality was a huge breakthrough. Data could now be stored at a central location, but it could be accessed from many different locations. The ultimate benefit of a NAS system is that it would allow users to collaborate more closely. Other vendors quickly started to include NAS functionality in their products.

1.4.3   Storage-area networks

For a period of time, NAS solutions solved the problem of allowing multiple users to access remotely stored data by many different users. However, problems remained. One of the biggest storage problems that firms faced was that they now had a large number of separate NAS systems. Whenever a new data set had to be stored or accessed, a new NAS was created, which grew quickly.

What companies needed was a way to pull together all of their separate NAS solutions into a single storage solution. This is why storage-area networks (SANs) were created. SANs helped organizations consolidate their storage assets to improve capacity utilization by sharing their storage resources effectively. The result of this was that companies were able to simplify storage management by using common software tools. SANs enabled the replication of critical information over long distances to provide greater levels of protection against data corruption and disaster events.

A SAN is a storage architecture that has been designed to connect detached computer storage devices, such as disk arrays, tape libraries, and optical jukeboxes, to servers such that the devices appear as local resources. SANs can then deliver storage to servers at a block level and provide feature mapping and security capabilities to ensure that only one server can access the allocated storage at any particular time. The protocol, or language, used to communicate between storage devices and servers in a SAN is SCSI.

The use of SANs helped companies consolidate their storage assets to improve storage capacity utilization by sharing their storage resources more effectively. This resulted in simplified storage management by allowing the use of common software tools and enabled replication of critical storage information over long distances to provide greater levels of protection against data corruption and disaster events. Many larger organizations, such as banks and telecommunication providers, were among the first to see the value in this and implement SANs.

SANs were not the perfect storage solution. Although the technology was powerful, it was the implementation that often caused a new set of storage-related problems for enterprises. One of these problems was that "islands" of SANs were being created. This was because there was little interoperability between SAN products created by different vendors and storage devices.

Even after a SAN was implemented by an enterprise, people were discovering that storage utilization levels were still fairly low. The reason for this was that traditional pre-SAN storage allocation methodologies were still being used. When this was coupled with the fact that companies could not replicate or move their storage between storage solutions provided by different storage vendors, this limited the storage options that were available to companies.

Storage virtualization is the process by which the physical storage of data is abstracted to create logical storage. The physical storage systems, disk drives, are first aggregated into "storage pools." These storage pools are then used to create the logical storage entities, which will then be presented to the applications that want to store and use data and which can be managed from a single console.

There are numerous reasons for virtualizing storage. These include being able to offer applications storage options that are more flexible along with simplified management of the stored data. As is the case with virtualized servers, once storage has been virtualized, better capacity utilization of the storage systems combined with better performance can be provided to the applications that use the virtual storage system.

A shared storage model has been created to show how applications actually store data onto physical storage systems (SNIA, n.d.). This model, consisting of four separate layers, is shown in Figure 1.10. At the bottom are the physical servers that will ultimately store the data. Sitting on top of that layer is the block aggregation layer, which contains the host, network, and device interfaces. On top of that is the file/record layer; at the very top is the application layer.

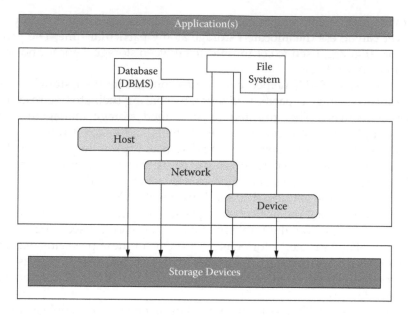

Figure 1.10 A shared storage model.

When a storage system is virtualized, the virtualization can occur at any layer. The layer at which the virtualization occurs then presents a virtualized view of the storage system to the layer above it. There are three different types of storage virtualization that can be implemented: server, storage network, and storage controller.

1.4.4 Server-based storage virtualization

Storage virtualization began in the OSs that run on servers. At a physical level, data is being stored on a hard disk in fixed-size blocks that could be read or written. However, keeping track of what data was located where on a given hard disk quickly became a challenge.

The OS implemented a rudimentary virtualization technique by which a file name was associated with a group of storage blocks. Now, by using a file name, an application could easily find the data that it wanted to reference. As the number of files grew, eventually the available storage on a given hard disk was exhausted, and additional hard disks (or "volumes") had to be connected to the computer to provide additional storage.

As with all things related to storage, soon the number of volumes that a server had to access to perform its tasks quickly unmanageable. OSs were then modified to support the idea of a logical volume manager (LVM). An LVM is simply a group of volumes that have been grouped together into storage pools (Figure 1.11).

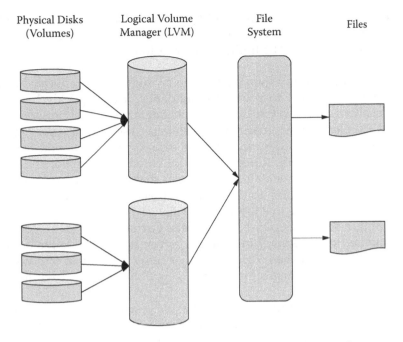

Physical Disks (Volumes)   Logical Volume Manager (LVM)   File System   Files

Figure 1.11 Logical volume managers used to store and retrieve information.

Because the application no longer had to worry about which volume it was retrieving its data from, this allowed the concept of "data striping" to be introduced to storage systems. When data striping is used, data that is being written out to a storage system is split between multiple volumes. Every disk has a built in delay, which occurs when it is asked to retrieve data while it performs a seek to find the start of the requested data on its storage platter. When the data is spread across multiple disks, this delay operation can be performed in parallel, and all of the data can be delivered in the time that it takes to read one segment.

Server-based storage provides three key benefits:

1. **Ability to stand alone**: No additional hardware has to be purchased to implement server-based storage virtualization. Any storage system that the server is able to access can be virtualized using this approach.
2. **Decreased cost**: The software required to implement server-based storage virtualization is already part of the OS used on the server. This means that no additional software has to be purchased or licensed.
3. **Flexibility**: Because server-based storage virtualization is part of the OS, it is easy to configure and is extremely flexible.

Using server-based storage virtualization does have its own set of limitations. These limitations may only affect certain configurations and usage circumstances. The limitations are as follows:

1. **Data migration**: When data has to be migrated or replicated across storage systems to provide redundancy, it can quickly become a challenge to keep track of the level of data protection provided.
2. **CPU slowing**: Virtualizing the storage system of a server takes a great deal of CPU processing. As data flows both to and from the server subsystem, it needs to be mirrored and striped, and parity bits need to be calculated. This means that computing resources will be taken away from the applications that are being processed on the server, resulting in a loss of performance.
3. **Single server**: Server-based storage virtualization is, by necessity, limited to working with a single server. It will maximize both the resilience and the efficiency of that single server but will not provide any advantages for other servers that need to access the same data set.
4. **Unique file systems**: The way that file systems are implemented on a storage system is dependent on the vendor that provided the storage. This means that for each storage system there is the possibility that a unique virtualization solution will have to be implemented and then maintained.

1.4.5   *Storage-network-based storage virtualization*

The arrival of SANs allowed the job of managing the storage and retrieval of data from a storage system to be moved off the server and onto special purpose-built hardware. Now, the servers communicated with the SAN system anytime they needed to store or retrieve data. It was not long before vendors realized that the SAN would be a good place to implement storage virtualization.

There are two types of storage-network-based storage virtualization: in-band and out-of-band, shown in Figure 1.12.

Both types of storage-network-based storage virtualization abstract the view of the physical storage systems between the server and the SAN. The only difference is if virtualization is performed in-band or out-of-band.

1.4.5.1   *In-band*

When in-band storage-network-based storage virtualization is being used, the virtualization function becomes part of the communication path that exists between the server and the SAN as shown in Figure 1.12. This functionality can either be built into a SAN switch or be provided by an additional piece of hardware.

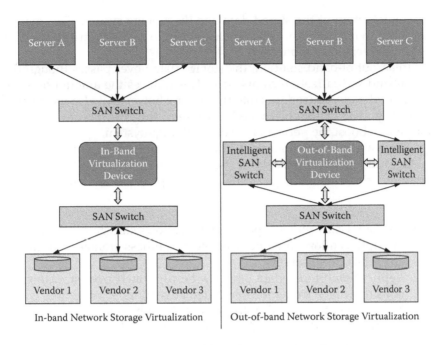

Figure 1.12 Types of storage-network-based storage virtualization.

In an in-band solution, the server will never see the physical storage system. It will only see the virtualization device/functionality. It is the responsibility of the in-band virtualization function to accept storage requests from the server, analyze them, perform a lookup on its storage-mapping tables to discover where the requested data is to be stored or read from, and then perform the storage operation.

A unique aspect of the in-band solution is that all of the storage data will flow through the in-band component. This means that it is possible to provide data on storage data usage, cache storage data, manage any replication services that are being used, and manage data migration operations.

1.4.5.2   Out-of-band

The difference between out-of-band solutions and in-band solutions is that in an out-of-band solution, the virtualization component does not lie in the path of the storage data. Instead, as shown in Figure 1.12, the virtualization component is connected to virtualization-enabled SAN switches that perform all of the required lookups.

Again, the servers have no direct contact with the storage system. Instead, the servers interface with the SAN switches, which are then responsible for interfacing with the physical storage system. In this type of a solution, the out-of-band virtualization component maintains a meta-data map of the stored data and the storage system. This map is then used to tell the SAN servers where to go to save or retrieve the data that the servers have requested.

In an out-of-band solution, the storage data never passes though the out-of-band virtualization component. This reduces the amount of delay that virtualization introduces into the system; however, it also means that the storage data cannot be cached. Because of this, the out-of-band solution cannot boost the performance of the storage system.

There are both advantages and disadvantages that the in-band and out-of-band solutions provide to the applications that use them. There are three primary advantages of using this type of storage virtualization solution:

1. **Pooling**: Using either type of solution allows storage solutions from multiple vendors to be pooled into a seamless storage pool that is accessible by all servers.
2. **Replication**: Stored data can now be replicated over storage solutions that come from multiple different vendors.
3. **Management**: The ability to manage the complete storage solution can be provided by a single management console.

The three most significant drawbacks are as follows:

1. **Complexity**: These are not easy storage solutions to set up. The physical storage elements have to be mapped to their virtual counterparts. This is done by creating a mapping table, which then becomes a potential single point of failure. The custom nature of these mapping tables means that vendors are able to lock in their customers once the table has been created. Moving data from a virtualized storage solution to a different storage solution can be difficult or impossible.
2. **Maintenance**: The virtualization device is generally a high-powered server that will require as much maintenance and as many software updates as any other server. The number of servers involved will grow as more are added to permit clustering solutions that will provide the needed backup and reliability that this part of the storage solution will demand. As the amount of data being stored grows, the servers may be hard pressed to scale with the solution.
3. **Latency**: In both solutions, the server's request to the storage solution no longer travels directly to the physical storage system. Instead, there is now a collection of SAN and virtualization devices in between the server and the storage system. This means that delay will be introduced to every interaction between the server and the storage system. Couple this with the limited

amount of CPU cycles and RAM that the SAN and virtualization devices have and there is the potential for delay to be introduced into the system.

1.4.6 *Storage-controller-based storage virtualization*

In a storage-controller-based storage virtualization solution, the vendor-provided storage array provides the virtualization services. This type of solution allows heterogeneous vendor storage arrays to be connected to the vendor's storage controller. Once this is done, all of the storage arrays are managed by the servers as though they were internal disk drives (Figure 1.13).

This type of solution can be implemented with a great deal less complexity than a storage-network-based storage virtualization solution. No additional layer of management software is required. The real value of this type of solution is that the storage controller is able to virtualize the storage; this allows each of the storage arrays to be used by the server as though it was a part of the storage solution no matter where the storage array may be physically located.

A storage-controller-based virtualization solution allows all of the functionality provided by the storage controller to be extended to all of

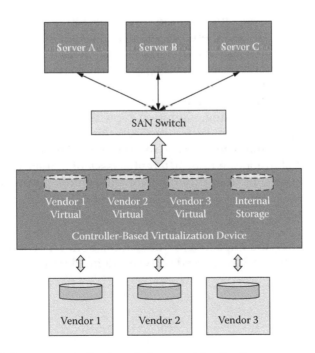

Figure 1.13 Storage-controller-based storage virtualization.

the storage components provided by all of the vendors. The real value of this is that data can be replicated between storage pools with no disruptions to availability of the data. In addition, replication can occur between storage systems provided by different vendors.

Storage resources can be allocated to specific applications. This means that specific cache, ports, and disk pools can be made available to a specific application. This can allow an application to achieve a given level of service or security policy.

In an enterprise storage solution, the use of a storage-controller-based storage virtualization solution provides more disaster recovery options. Clustering capability functionality between different controllers has been created to permit the real-time recovery from the failure of a complete storage array failure.

Storage-controller-based storage virtualization provides five clear benefits to users:

1. **Standardization**: The way that servers are connected to external storage pools is performed via industry standard protocols. This means that users can avoid any type of proprietary vendor lock-in.
2. **Simplicity**: This solution allows reduction of the amount of SAN hardware required. This means that the complexity of the solution is also reduced.
3. **Decreased cost**: This solution builds on top of the existing SAN infrastructure and therefore often is cheaper than other storage virtualization solutions. In addition, the tools required to manage the storage solution (availability, replication, management, etc.) can be consolidated.
4. **Legacy support**: The most important storage system functions, including provisioning, migration, replication, and partitioning, can now be performed on legacy storage equipment from other vendors.
5. **Data protection**: The ability to replicate stored data between storage pools that are located on different classes of storage equipment or even equipment from different storage vendors means that overall data protection costs can be lowered and data protection solutions expanded.

In summary, the impact of virtualization, both on memory management and on servers, has enabled software-defined networking (SDN).

chapter two

Software-defined networking

2.1 Introduction

Software-defined networking (SDN) refers to a way of organizing computer network functionality. SDN allows the network to be virtualized, providing greater control and support for traffic engineering.

2.2 Network limitations

Current operating networks are the result of protocol and network design decisions initially made in the 1970s. At that time, it was envisioned that 32 bits would be sufficient to handle all of the Internet Protocol (IP) addresses that the Internet would ever need (roughly 16 million)—and it was, until February 2011, when the IP4 addresses ran out (Cert, 2013). Networks were also considered to be static. Once established, a network topology was not expected to change much, if at all.

The servers that connect to today's networks have undergone a dramatic transformation in the past decade. The arrival of server virtualization has fundamentally changed the role of a server. Severs are now dynamic and can be created and moved easily, with an increase in the number of servers that are able to use the network. Before the arrival of large-scale virtualization of servers, applications were associated with a single server that had a fixed location on the network. These applications used the network to exchange information with other applications that were similarly in fixed locations.

This has changed. Applications can be distributed across multiple virtual machines (VMs). Each of these VMs can exchange traffic flows with the others. Network managers can move VMs to optimize and rebalance server workloads. Application movement can cause the physical end points of an existing flow to change. The ability to migrate VMs creates challenges for many aspects of so-called traditional networking, including addressing schemes and namespaces along with the basic notion of a segmented, routing-based design (Open Networking Foundation, 2012).

Along with the virtualization of servers, many companies are also using a single network to deliver all of the voice, video, and data networking needs that they have. In today's legacy networks, the concept of quality of service (QoS) is used to provide a differentiated level of service for different applications. However, the provisioning of many QoS tools is highly manual. Network staff must configure each vendor's equipment separately and adjust parameters such as network bandwidth and QoS on a per session, per application basis (Open Networking Foundation, 2012). Legacy networks are static, and the networks cannot dynamically adapt to changing traffic, application, and user demands.

Software-defined networking emerged from research work initially performed in 2004 as part of a search for a new network management paradigm, which resulted in two different forms: the routing control platform (RCP 40) work that was done at Princeton and Carnegie Mellon University (Caesar et al., 2005) and the network security work done at the same time as a part of the SANE Ethene project at Stanford University and the University of California at Berkeley (Casado et al., 2006).

This initial work was built on in 2008 by two different groups. The startup Nicira, which was eventually bought by VMWare, created NOX, a network operating system (Gude et al., 2008). At the same time, Nicira worked with teams at Stanford University to create the OpenFlow switch interface (Ferland, 2012; Open Networking Foundation, 2013).

In 2011, the de facto standards body of the SDN space, the Open Networking Foundation (http://www.opennetworking.org), already had broad networking industry support. Membership included firms such as Cisco, Juniper Networks, Hewlett-Packard, Dell, Broadcom, and IBM. Its board consisted of members who were drawn from some of the networking industry's biggest firms: Google, Verizon, Yahoo, Microsoft, Deutsche Telekom, Facebook, and NTT.

One of the most significant events in the history of the SDN technology occurred in 2012. At the Open Networking Summit, which was attended by over 1,000 engineers from the networking industry, Google made the announcement that they were already using SDN technology. Google said that they use SDN as a part of the wide-area network (WAN) used to interconnect their data centers (Hölzle, 2012).

Today's networks are built by interconnecting tens, hundreds, and sometimes thousands of sophisticated network routers whose sole purpose is to accept data packets from applications and then to forward them to the next router on the path to the data packet's eventual end destination.

Figure 2.1 shows a high-level view of the components of a modern network router. A router, as provided by a vendor, creates a unified stack of functions that work together. This functionality starts with the specialized packet-forwarding hardware whose job it is to accept data packets and then forward them to the next router to eventually deliver the packets to the destination application. Controlling all of this is the router's operating system. The operating system is the router vendor's "crown jewel" and has been optimized to work with the vendor's underlying hardware platform. On top of this are the various network applications,

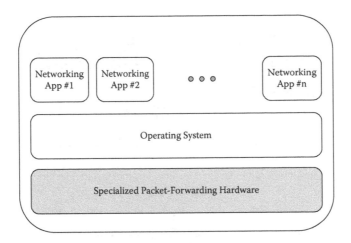

Figure 2.1 Architecture of a modern network router.

such as specialized routing protocols that use the operating system and the packet-forwarding hardware to accomplish the routing objectives.

To make changes in network behavior, it is necessary to access each router in the network and issue a set of operating system commands in the language that has been defined by the router vendor. The effect of this is to change router behavior. This is a closed environment, and routers cannot easily interact with other components that compose the network.

SDN research was motivated by current problems in computer networking. Computer networking today uses three separate planes to accomplish tasks: the data plane, the control plane, and the management plane (Figure 2.2). The data plane is tasked with the job of processing the packets received. When a packet arrives, the data plane functionality uses its information about the local forwarding state and the information that is contained in each packet's header to make a decision about whether to drop the packet or forward it. If the data plane decides to forward it, then it must decide which computer to send it to and which port on that computer packet should receive it. To keep pace with all of the packets that are arriving, the data plane processing must be done extremely quickly.

The control plane computes the forwarding state that the data plane uses to forward the packet. This forwarding state can be calculated using distributed algorithms or centralized algorithms, or it can be manually configured. The data plane and the control plane are very different. They have both been designed to accomplish different jobs. The management plane is responsible for coordinating the interaction between the control plane and the data plane.

Figure 2.3 shows the way that the data plane has been constructed using abstractions. An unreliable set of transport layers has been used ultimately to create a reliable transport layer.

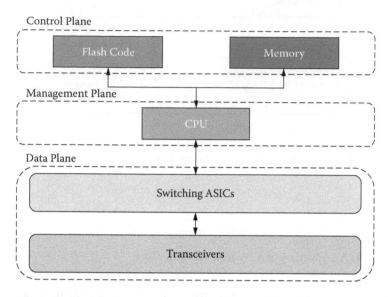

Figure 2.2 Three planes of a traditional router.

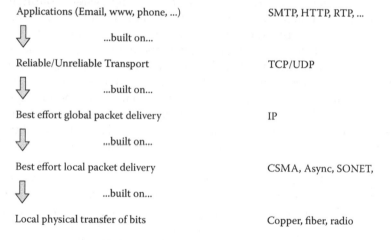

Figure 2.3 Data plane abstraction layers.

No similar set of abstractions exists for the control plane. There is no underlying simplification that has been used to build a control plane. Instead, a large number of mechanisms have been created to implement different types of control planes.

The control plane mechanisms have all been designed to accomplish a variety of different goals. One example is the routing control planes that implement a wide variety of distributed routing algorithms. There are also isolation control planes that can be used to provide access control lists (ACLs), virtual local-area networks (VLAN), firewalls, and so on. Finally, there are traffic-engineering-based control planes that can use adjusting link weights to make routing decisions, implement Multiprotocol Label Switching (MPLS), and the like. All of these different approaches are trying to affect the routing of packets—they want to control how the forwarding state is calculated.

The problem with the control planes that are used in today's networks is that they are not modular—they cannot be used together. Each one of the control planes solves part of the problem of controlling a network, but none of them solves all of the problems of controlling a network. Effectively, each provides limited functionality.

The modern network control plane provides far too many mechanisms but with little functionality. This is because there has been no way to abstract exactly what the control plane does. Every time a new problem is encountered (e.g., people are illegally gaining access to a network), a new solution is defined. All of these new solutions only implement some of the functionality that is required. No single solution has all of the required control plan functionality.

The result of designing the control plane this way is that a disharmony is created between the enterprise network and the enterprise's business needs. A simple example of this would be data route selection in today's legacy networks. If an application has a great deal of data that needs to be transferred to another application, it may end up using the fastest and most costly link in the network simply because it is available. However, if the application were able to "talk" to the network, it might discover that there was a slower and less-expensive link that would serve its purposes just as well. However, the design of the control plane used in today's networks does not allow this type of information to be shared.

2.3   Network control plane

The control plane is designed to compute the forwarding state under three different constraints:

1. It has to be consistent with the low-level hardware and software. The ASIC (Application-Specific Integrated Circuit) must be known and available at the CLI (command line interface) to know what the hardware and the software are doing in the switch.
2. It must be based on the entire network topology.
3. The forwarding state has to be inserted into every single physical forwarding box in the network.

Every time a new network protocol is designed, these three problems are revisited and another solution is created. This is not a sustainable approach.

Programming analogy

Pretend a programmer has been asked to write a program. The program has the following requirements: (1) It has to be aware of the hardware on which it is running (registers, limited available operations, etc.), and (2) it must specify where each bit would be stored.

A programmer would not be willing to deal with this type of low-level complexity for long. Abstractions of the hardware system would be created quickly. First, a programming language that was independent of the physical hardware would be created (compiler); then, a virtual memory interface (operating system) would be created. Programmers have learned to use abstractions to separate the solutions that they create from the real-world concerns of the systems on which their solutions are being implemented. This is exactly what network designers should do also.

Figure 2.4 shows the different architecture that a router would have in SDN. The data plane functionality would be all that the router hardware would provide. The control plane and the management plane would be provided by a software application that was executing on a separate platform connected to the router via a secure data link.

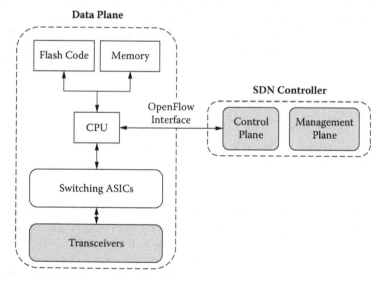

Figure 2.4 Router components in SDN.

To define functions for the network's control plane, a specific function set is needed. The first is a type of general forwarding model that can hide the details of the low-level hardware and software that has been used to create the network switch. Next, a function to determine the current network state is needed. This would be used to allow decisions to be made based on the entire complicated network. Finally, a function to configure the network is needed. The goal is to avoid having to configure every physical box in a network as there could be thousands or even millions of boxes in a single network. Therefore, a function is needed to simplify configuration. This can be done by computing the configuration of every physical device.

2.3.1   Forwarding function

When thinking about the best way to design a network control plane, one of the first functions to implement is the packet-forwarding function. Ignoring how this functionality has been done in the network switches that are deployed in today's existing networks, it is clear that the forwarding function needs to be implemented independently of how the network switch is actually implemented. The goal is to be able to express what should happen to a packet without having to worry about which switch is going to be used to implement it. This means that the network switch should be able to use any set of ASICs with varying degrees of capabilities, and this should have no effect on the forwarding function. In addition, the software that is executing on the network switch could be from any vendor, and this also would have no impact on the forwarding function.

In the constantly evolving world of SDN, the OpenFlow interface is one proposal for communication between centralized control plane software and the network switch. Vendors such as Cisco and VMWare have also developed their own alternative proposals. The OpenFlow interface is a set of application programming interfaces (APIs) that would allow an external software application to communicate with a network routing switch. The network packet that the APIs would be discussing is called a flow entry. A flow entry takes the form of a packet template and an action and is defined like this: <header, action>. A flow entry says that if a packet matches the template, then the switch should take the action (e.g., drop, forward out a specific port, etc.). OpenFlow is simply a general language that every switch in the network has to understand.

At a high level, the idea of an OpenFlow interface is easy. However, when the details start to be examined, things become considerably more complex. To implement OpenFlow, designers must make many different decisions. These can include how best to perform rapid header matching or what actions are going to be allowed, once a header has been matched.

2.3.2   Network state function

To create a network state, the first thing is to find a way to "abstract away" all of the complicated distributed functionality that is going to be required to collect the information that is necessary to create the network state. The ultimate goal of the network state function is to present a "global network view." This looks like a graph (objects and links) that has network information associated with it: network delay, link capacity, recent loss rate. Once the network graph is created, then the controlling software is able to make decisions about what to do based on the network graph. If access to the graph is provided via an API, then the actual network elements that make up the network could be controlled via the API.

The global network view functionality could be implemented as part of a network operating system. This software would run on servers that are separate from the switches that are used in the network. It could, of course, be replicated to increase its reliability.

To keep all of the network data current, information would have to flow bidirectionally between the network operating system software and the network servers. This information flow would allow the global network view to be constantly updated with a view of what was happening at each switch in the network. Likewise, each switch in the network could then be updated to provide accurate control of packet forwarding.

Figure 2.5 shows a typical network, using traditional network design. This network consists of a set of network switches that have the job of routing packets of user information between them. Connections exist between some of the switches in the network, but not all. Every switch is reachable via one or more paths.

These connections are also used to allow each of the network switches to exchange control information that they can then use to update their individual "view" of the network. This information is then used to modify how the switches forward the received packets.

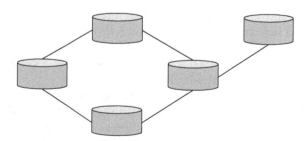

Figure 2.5 Network of switches.

Figure 2.6 shows an updated version of the network shown in Figure 2.5. More detail has been added on the control mechanisms that are being used in this network. Standards are being used to implement a peer-based routing algorithm. This algorithm is, by necessity, a distributed algorithm that runs between neighbor switches in the network.

Because of the way that networks are designed and implemented today, this control algorithm is both complicated and task specific. As an example, the control mechanism could be a shortest path first (SPF) algorithm that has been customized to allow the calculation of two disjoint paths that will be used for packet forwarding.

Figure 2.7 shows how today's complicated network design can be simplified by the new approach that SDN offers. In this network, there is a general-purpose software algorithm that is now running on the network operating system servers. This software talks to each of the switches in the network to determine the topology of the network.

Distributed Algorithm Running Between Neighbors
Complicated Task-Specific Distributed Algorithm

e.g., SPF routing algorithm

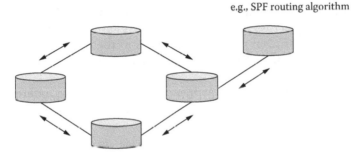

Figure 2.6 Traditional control mechanisms.

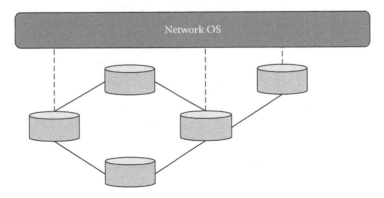

Figure 2.7 Software-defined network (SDN).

Once this information has been collected, the network operating system software is then able to create a global network view, as shown in Figure 2.8. The key characteristics of this piece of software is that it is both extensible and flexible.

Once a global network view is available, it is then possible to create a control program that operates at a higher layer. The control program can take on many different forms: It could be a routing program, an access control program, a traffic engineering program, and so on. All of these different types of control programs can use the global network view to make packet-forwarding decisions for the network. Note that this centralized decision making based on a global state is different from today's networks, in which distributed decisions are made based on imperfect local knowledge of the network's global state.

Network control is regarded differently in SDN. When SDN is used to control a network, how the control program operates is much different from how routing is managed in a traditional network. The way that the information is used to determine how each of the switches in the network should forward packets is now going to be determined based on the global network view.

The method the network uses to determine how to forward packets that are traveling across it has been radically changed. Instead of using a distributed algorithm in which each switch in the network does its own set of calculations, now there is a centralized control program that has that responsibility.

This control logic runs as a program on the network operating system and uses an API to interact with the global network view. Ultimately,

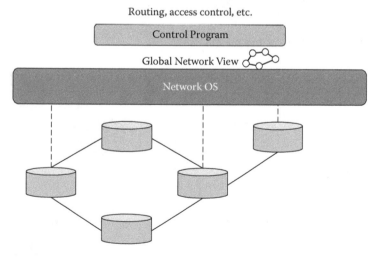

Figure 2.8 SDN with global network view and control program.

the best way to envision the control program is to view it as a type of graph algorithm.

This changes how things will be done when there is a need to do something different with the network in terms of how packets are routed. Instead of having to redesign a distributed routing algorithm, only the centralized control program will have to be modified. This makes the way that packets are forwarded much easier to verify, maintain, and investigate.

2.3.3   Configuration function

The control program is where the packet behavior is going to be expressed. The underlying network by itself will have no idea how packets are to be routed. As an example, if it is determined that packets coming from node A should never travel though node B, then the control program will be responsible for expressing this. The underlying network would never have a way of telling if packets from A should or should not be routed through node B.

The control program will not be responsible for implementing the routing behavior on a physical network infrastructure. In the example, the control program will not be responsible for implementing the rule that node A should not talk to node B. Instead, that will require detailed configuration information be placed in the forwarding tables in every router along every path in the network.

To implement this separation between the control program and the implementation of the output of the control program, in SDN the control program interacts with an abstraction of the underlying network in which many of the details have been removed.

The global network view graph with which the control program would interact only has enough detail so that the control program will be able to express its goals ("Node A's packets will never travel through node B"). The amount of detail that the global network view will provide to the control program will depend on the function that the control program is trying to do. QoS, access control, or even traffic engineering would all require different amounts of detail on the underlying network.

A way of thinking about this concept is to relate it to the world of programming languages and compilers. The computer that will eventually be used to run the program has a specific instruction set, and eventually the program that the programmer writes will have to be expressed in that language. However, when the programmer is writing the program, the level of detail that goes along with writing a program in the computer's native language is not needed. Instead, a programmer can use a higher-level language and not have to worry about all of the low-level details of how to get the program to run on the actual hardware—the compiler will take care of this.

As an example, Figure 2.9 shows a sample network. Assuming that a control program is executing an access control function, it will want to prevent node A from communicating with node B. In a traditional network, the control program would have to perform the computations to determine how routing would be permitted to be performed in this network, and then it would have to go and configure the routing tables on every node in the network to implement the results of that routing algorithm.

An alternative way of implementing the access control function is to present the algorithm with the view of the network shown in Figure 2.10. In this figure, only the external interfaces on the network are shown to the algorithm.

The access control algorithm is asked the question, "Which nodes should be allowed to talk to which nodes?" Once the access control functions identify that node A should not be able to route packets through node B, then its task is complete.

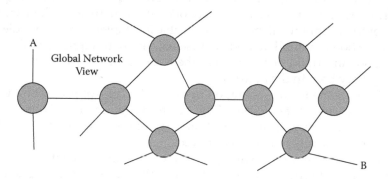

Figure 2.9 SDN example: access control.

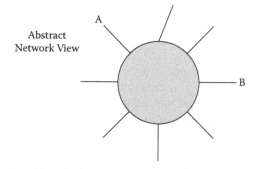

Figure 2.10 SDN example: access control abstract view.

It will now be the responsibility of the "compiler" to make sure that all of the correct packet-routing information is placed into the correct packet-forwarding tables in each of the switches that make up the network. To allow the compiling of the control program's packet-routing decisions to be translated into actual configuration commands for each of the switches in the physical network, a new layer has to be added to the SDN model. This layer is called the virtualization layer (Figure 2.11).

One of the jobs of the virtualization layer is to present an abstract network view to the control program—a view that has all of the unnecessary details removed. The control program will "see" a simple view of the network. In the example, the control program will then decide that it wants traffic from node A not to be routed through node B.

The virtualization layer will take the access control decisions made by the control program and convert the decisions to the global network view. The updated global network view is then handed off to the network operating system, which then has the responsibility for properly configuring each of the switches that make up the actual physical network.

One additional point to make here is that the underlying hardware no longer has to be a sophisticated router. Instead, a relatively simple device that can provide basic packet-forwarding hardware will do. This simple device will communicate with the network operating system using a protocol such as OpenFlow.

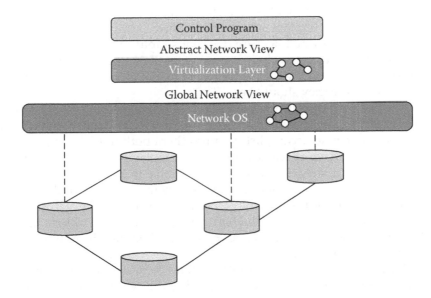

Figure 2.11 Components of an SDN.

2.3.4   Separation of functionality

The architecture of an SDN-based network has a clean separation of functionality. The control program portion of the network has one job to do: express the operator's requirements for how packets are to be routed in the network. The functionality of the control program is driven by operator requirements. When operator requirements change, the control program is the only portion of the SDN environment that has to be changed.

The virtualization layer is built to take the abstract view that is presented to it by the control program and translate it into the more detailed global view. It is able to do this because its actions are driven by the specification abstraction for the particular task that it is doing.

Finally, the network operating system takes the packet-routing control commands that have been created based on the global network view and maps them to the physical switches that make up the network that is being controlled. The API that the network operating system uses to talk to each of the network switches is driven by the network state abstraction. The interface between the network operating system and the switches is driven by the forwarding abstraction—potentially the OpenFlow interface.

Note that if there are changes in the underlying physical network, then these changes will travel up to the network operating system. Once there, they will be reflected in the global network view and eventually in the abstract network view. Working together, the network operating system, the virtualization layer, and the control program make up the layers of the SDN control plane. Just as the network data plane has always had layers to implement its functionality, in an SDN-based network, now the control plane has layers to implement its functionality.

The use of SDN architecture does not simplify the task of creating a properly operating network. However, what SDN does do is to take the complexity associated with this task and move it around. The hard parts are placed where they should be.

The network operating system and the virtualization layer components of SDN remain complex pieces of software. The functionality that these layers have to accomplish requires them both to be large pieces of software and to have a great deal of logical complexity designed into them.

However, an SDN designed network is able to accomplish two main objectives. The first is that it is able to simplify the interface for the control program portion of the control plane. This portion of the network can now be made user specific. The hard part of creating a properly operating network, the complex code, can now be pushed into the reusable SDN platform piece of the design.

Note that this design is similar to how compilers for software languages are designed. Compilers are hard to create. They contain a great deal of complexity, and they have to have a good understanding of the physical computer on which the program will eventually be running. However, once a correctly working compiler has been created, then the hard work has been done, and programmers can just worry about stating what they want to have happen and not how it will happen.

2.4 Applications

In SDN, the server that is executing the control plane will also be able to run a set of additional network applications. Many of these applications are executed on individual routers in today's legacy networks. The applications will include such functionality as packet forwarding, virtual private networks (VPNs), security, bandwidth, network virtualization, load balancing, and so on. However, none of these applications is important enough to cause the industry to shift to using SDN-based networks.

In the computer networks that are being designed and implemented today, topology is policy. Where routers and firewalls are placed in the network will limit where broadcast domains can be, access control lists, and so on.

Many enterprises are now considering moving their corporate networks from the world of the physical domain to the world of cloud computing. Having implemented a topology that allowed them to create the corporate network policy that they wanted, these enterprises would like to keep the same policy for their cloud-based networks.

The problem that these enterprise network operators are now encountering is that few of them have an abstract view of what their corporate network policy is. What they are missing is the so-called network algebra that describes who can (or cannot) talk with whom.

The true power of using an SDN is that the network operator can now specify a "virtual topology" of their enterprise network to the cloud. The cloud network, which has been implemented using SDN technology, can now ignore the physical design of the network that it is replacing and instead implement this policy. The end result is that the enterprise can now migrate seamlessly from or to the cloud from their current network.

To make this happen, the correct network policy has to be embedded into the cloud. The current network policy has to be "read" out of the existing network. Network operators can use their current network topology to create their network policy statement. This policy statement can then be replicated in the SDN-enabled cloud. This allows, for example, VMs in the existing network to be moved to the cloud, and the same policy requirements will apply because it will exist in the same virtual topology.

SDN technology gives enterprise network designers and operators the ability to virtualize their networks. No other technology gives them this ability. This is why network virtualization is the compelling enhancement of SDN over traditional networks, the "killer app" of SDN.

chapter three

SDN implementation

3.1 Introduction

Software-defined networking (SDN) can be regarded as a set of abstractions governing how the control plane in a modern network is specified. SDN is not a set of mechanisms, as SDN can be implemented in a number of different ways. Although much of the discussion about SDN involves the OpenFlow interface to the physical network's switches, OpenFlow should be thought of from a technical viewpoint as the least-interesting component of SDN.

Unlike today's distributed network routing protocols, SDN can be thought of as simply computing a function. SDN computes functions on an abstract view of the underlying physical network. This allows SDN to ignore the detailed physical infrastructure that has been used to implement the actual network and permits network control plane engineers to manage network traffic without being constrained by the physical network design. In SDN, the network operating system (NOS) is responsible for taking a specified function and making sure that the results of the function are distributed to every switch in the network.

Ultimately, this leads to SDN's so-called killer app, which is its ability to allow physical networks to be virtualized. In modern computer networks, the servers and the storage functions have already been virtualized. Virtualizing the network itself is simply the next step. With the virtualization of the network, the last stage in network designers freeing themselves from physical reality has been achieved.

Once the physical network has been virtualized, SDN gives the software applications that are using the network the ability to reconfigure the network to suit their current needs. This ability allows the network to provide optimal service to the applications that are using the network.

3.2 SDN design

One of the immediate benefits to creating networks based on SDN concepts is simplifying network management. The widespread adoption of SDN technology has the possibility to change the way that networks are both designed and built.

43

3.2.1   Separation of the control and data planes

The initial impact is that, for the first time, the network's control and data planes can be separated. In non-SDN environments, the control plane and the data plane are tied closely together. The network switches that compute the routing tables are the same devices that then implement the routing tables. A side effect of this is that both the control plane and the data plane are currently provided by the same vendors.

SDN completely changes this. The control plane and the data plane are pulled apart. The control program can run on one set of servers, and the NOS can run on a completely different set of servers. The NOS will observe and control the data plane; however, it is not part of the data plane.

One of the greatest potential impacts of implementing SDN is that it presents the possibility that the computer networking industry is going to be fundamentally changed. In a network that is built on SDN technology, the enterprise network designer can now purchase the control plane from third-party vendors, which can be done independently from the vendors that provide the switches. Because the "network intelligence" has now been removed from the switches and resides in the SDN layers, the network switches have been transformed into commodity hardware.

The implementation of SDN-based networks will also cause fundamental changes to how network testing is performed. Currently, the only way to test a network is to build a mirror image "test network" with the same type of switches planned to be deployed in the "real" network. Network engineers prepare test scripts with anticipated network configurations and test the configurations to see how they perform and how well implemented the network profile is in the target environment.

In SDN, this approach changes. As all of the network hardware exists behind a common interface, it is easy to conduct unit testing of the hardware to ensure that the interface is working correctly. Once this testing is completed, because the control plane now exists in software, large-scale simulations of the control plane are possible. The ability to test a network design before placing it into production was one of the reasons put forth by Google when the company explained why it had implemented a SDN as part of their interdata center wide-area network (WAN) (Neagle, 2012).

3.2.2   Edge-oriented networking

In SDN, the majority of important networking functionality can be done at the edge of the network instead of within the network. This functionality includes such things as quality of service (QoS), mobility, access control, migration, network monitoring, and so on. This changes the use of the core of the network. The network core will be used simply to deliver network packets from edge to edge. The network protocols that are used in today's network do a very good job of performing this function.

Pushing the network functionality to the edge of the network has two important implications for how SDN will be designed in the future. The complexity associated with SDN can be pushed out to the edge of the network. This is where all of the complicated functionality associated with matching packets can be performed. Effectively what is being done is that an "overlay network" is run on the core of the network.

Therefore, SDN will now be incrementally deployable. The network can be divided into two types of switches. The first is the core switches, which will be legacy switches—the ones that are used in the network today. The edge of the network can become a software-only switch that runs on a server. This switch can run a program that simulates an OpenFlow switch, and this can all run on the Linux operating system, which in turn is running on the Xen virtual machine hypervisor.

What this means is that an SDN can be deployed without the need to make any changes to the core network. The virtual edge switches can be added to an existing network, and they can tunnel across the legacy core physical switches. It is entirely possible that none of the legacy hardware switches may ever need to support the OpenFlow interface.

The other big change is that the SDN offers the possibility that the network can become software oriented. The complicated parts, such as calculations around how to forward a network packet, can be done in software that runs at the edge of the network. The network's control plane is now a program that can run on almost any server. This is a significant change from having the control plane be part of a closed proprietary switch or router box that is running a protocol.

Network creation will also change. No longer will networks be designed. Rather, they will be programmed. The focus of the "network programmers" will be on the modularity and the abstractions of the underlying physical network. The focus will no longer be on packet headers. The true value of this is that networks will now be able to innovate at software speeds and will not be limited to innovating at hardware speeds.

Software by its very nature lends itself to clean abstractions. This means that networking can now become a much more formal discipline. The clean abstractions and separations between layers in SDN will allow increased rigor.

An example of this would be how WAN control would be designed in SDN. One of the big challenges in this type of network has to do with link failure. When link failure occurs, all of the routers in the network need to have their routing tables updated. The official way to describe this is to say that the routers need to "converge." In the example here, convergence is defined to be the state of a set of routers that have the same topological information about the WAN in which they operate. To be able to say that a set of routers has converged, they must have collected all available topology about the WAN from each other via the WAN's implemented routing protocol, the information they gathered must not contradict any other router's WAN topology information in the set, and it must reflect the real state of the WAN. Another way of saying this is that, in a converged network, all routers must "agree" on what the WAN topology looks like.

In SDN, because the NOS will update all of the routing tables, there will be no iterative convergence for the routers in the WAN. This means that there will be a bounded depth of computation—the update will take a specified amount of time, not an indeterminate amount of time.

Looping is a significant problem in today's network. Looping can occur when there is a transactional update to some of the network routing tables while a network packet is making its way across the network. The packet that was sent when the network was in state A now finds itself being routed in a network that is in state B, and looping can occur. Because SDN updates the abstract and global network views at the same time, there will be no network disruption during convergence, and looping will not occur. This is something that cannot be said about most modern routing algorithms.

One other benefit of SDN's clean abstraction is that network trouble-shooting now becomes easier to do. When an operation that is not supposed to be occurring in the network is detected (packets from node A are being routed through node B), beginning at the control program level, first network engineers will check to see if this behavior is ever supposed to happen. If the answer is no, then the engineers check the flow entries at each level in the abstraction. The goal is to identify the first place where the packet forwarding rules were broken.

Because the SDN control plane is implemented in software, once the location of the problem has been identified, a test program can be run. By running a simulation, the minimal casual set that can lead to the breaking of the rule can be identified. This means that network problems can now be detected (and solved) algorithmically.

The greatest benefit of SDN is not better overall data center behavior. Rather, the greatest benefit that SDN will provide is that it will become easier to reason about how network control is going to behave. When a network is built using SDN technology, the resulting network will be different from today's networks. The hardware that is used to build the network will be cheap, it will be interchangeable, and it will allow networks to follow Moore's law. The software used in SDN can be updated frequently through software releases, and the software will not be dependent on the network hardware.

Finally, the functionality of SDN will be driven mainly by software. This software will reside at the edge of the network in software switches and in a control program. The entire network will be built on a solid intellectual foundation.

3.3 SDN operation

Networks that are built using SDN technology operate differently from today's legacy networks. To show how SDN might operate, Figure 3.1 shows a sample network that has been constructed using four switches. Various servers are connected to each router and different parts of a given application run on different servers.

The routers have been connected to create a mesh network. In a traditional networking environment, the routers would automatically exchange information, and by using the spanning tree algorithm, they would quickly discover that this network contains the possibility of an infinite loop (Ferland, 2012). Various networking ports would be configured to be placed into blocking mode so that packet-forwarding problems associated with loops would be avoided.

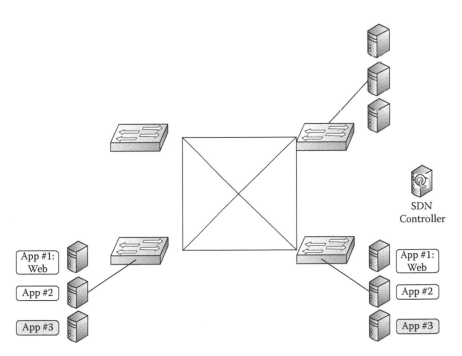

Figure 3.1 Sample SDN environment.

However, because this is SDN, none of those actions will occur. Instead, the SDN controller that is shown on the right side of Figure 3.1 will be responsible for providing all of the routers with the data needed to populate their routing tables. Instead of concerning itself with network-level details, the SDN controller will be instructed to permit the web front-end portion of App 1 to be able to talk with the database back-end portion of App 1, which is located on a different server. What are actually being defined on the SDN controller are the application data flows.

Figure 3.2 shows the next step in the process: The SDN controller updates each of the routers in the network using the OpenFlow protocol with the flow table contents that the controller created based on the application data flows. This is unlike traditional networks, where the router flow tables would have been based on such things as virtual local-area networks (VLANs), spanning trees, routing protocols, subnet destination addresses, and so on.

Figure 3.3 shows an application being moved to a different server in the network. When this occurs, the SDN controller will be informed, and it will then send updated flow table contents to each router so that the front and back ends of App 1 are still able to communicate with each other.

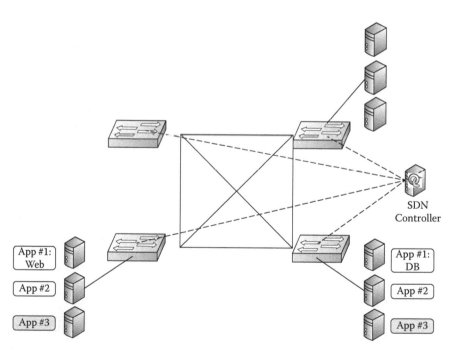

Figure 3.2 SDN controller updating routers.

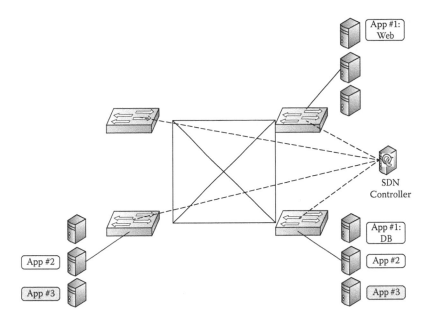

Figure 3.3 Application movement in an SDN environment.

In this SDN configuration, the network applications have the ability to talk directly with the SDN controller. Situations in which they might want to do this include if an application is going off line and no longer needs the network bandwidth that has been reserved for it or an e-commerce application is unable to process any additional orders and requests that orders no longer be sent to it. When the SDN controller receives these types of notifications, it will again refresh the flow table contents on each router so that the requested network behavior is implemented.

chapter four

Service providers and SDN

4.1 Introduction

Telecommunication service providers have constructed some of the largest and most complicated networks currently in existence. Now, they are facing a severe problem as revenues are incrementally increasing and network equipment costs are growing exponentially (Figure 4.1). For these service providers, their interest in software-defined networking (SDN) is all about economics.

The cost of sophisticated networking equipment continues to increase as vendors add more and more functionality to their boxes. Unacceptably high network equipment cost escalations will result in a nonsustainable business case for the service providers. At current rates, it is entirely possible that the service provider's network revenues will eventually be severely diminished by the cost of providing the network. Service providers have become interested in SDN because they believe that SDN can lower their cost of building and operating networks.

Service providers want to avoid building networks composed of purpose-built hardware that is sold to a relatively few service provider customers (Elby, 2011). Instead, service providers envision a network where the underlying hardware is commercial off-the-shelf (COTS) hardware that follows mass-market cost curves and will result in the ability of the service providers to lower their equipment expenditures. The ultimate goal of the service provider is to have the cost of building the network increase at a rate that most closely matches the rate at which network revenue is increasing.

To make this happen, every service provider would like to be able to match network expenses to revenue growth. This would mean that instead of having to build an expensive network and hope that customers will show up and use the network, service providers could build out the network based on the revenues generated by the customers who are currently using the network. If they could achieve this, then the service providers would be able to report consistent profit margins.

Service providers see SDN as perhaps finally providing them with the tools that they need to enable inexpensive network feature insertion. The belief is that these new features will result in a boost in revenues. Service providers realize that there is always a risk that creation of these

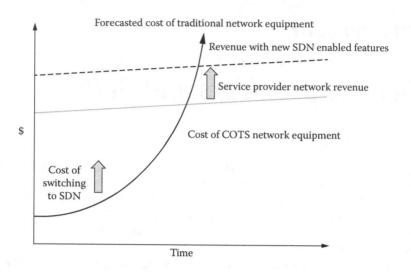

Figure 4.1 Telecom service provider's SDN motivation.

features may never happen, so they are not basing their SDN business case solely on this occurring.

All of the benefits of switching to SDN have to be counterbalanced against the cost of creating an SDN environment. The cost of switching from a legacy network to SDN cannot overwhelm the benefits. The cost must be balanced by the amount spent and how quickly the SDN environment is created.

4.2   Telecommunication SDN attributes

For SDN to be a successful part of the networks over which telecom service providers offer service, the network will have to have several different attributes. These attributes have been identified based on the past experience of the service provider with legacy networks.

The first needed attribute is for SDN to have a network operating system that supports a service- or application-oriented namespace (Elby, 2011). Telecommunication service providers have realized, with their current networks, that they have been spending far too much time and have been purchasing too much software and hardware to get their networks to work at the packet level. Much of what happens in an existing network has to do with Internet Protocol (IP) packet addresses and attachment points. The service providers would prefer to be interacting with their networks in terms of service-specific information, such as policy and service routing. These service features do not necessarily use a packet's header information, and it can be difficult to manage an existing network at this level.

Another key attribute of SDN is the ability to virtualize as many of the network resources as possible. Service providers are motivated to monetize all parts of the network that they build. This means that resource virtualization is a critical SDN need. If SDN can provide the virtualization feature, then service providers will be able to create networks that support multitenancy. This means that, just like virtualized servers that can run multiple operating systems at the same time, the network that has been built will be available to support the provider's services as well as those of the other service providers or enterprise users with access. Network virtualization will provide the elasticity and the aggregation that will be necessary to allow the service providers to pool resources to achieve scaling.

Service providers are looking to simplify how they interact with the network. To do this, they believe that the various components of the network have to be separated. This means that the topology, the traffic, and the interlayer dependencies all have to be decoupled. In today's networks, all of these items are closely tied together (e.g., in an IP Multiprotocol Label Switching [MPLS] network). Things such as IP forwarding or the services that are being provided are closely tied to the underlying topology. The ability to separate these layers gives service providers a powerful motivation to move to SDN-based networks.

Finally, a critical attribute of SDN from a service provider's point of view is that SDN will support the careful introduction of the new SDN architecture to interwork with the large existing base of legacy networks when the new functions will provide the greatest value. This is where potential SDN components such as the OpenFlow (Azodolmolky, 2013; Open Networking Foundation, 2013) control interface (along with complementary management protocols) will be used to enable new types of control paradigms on existing legacy network equipment.

Figure 4.2 shows one possible deployment scenario for carriers introducing SDN technology to their networks. The left side of this picture captures the end user, either a consumer with access to a high-speed Internet connection or an enterprise with optical access. Note that the end users could also be accessing the network using high-speed wireless technologies such as 4G LTE.

In Figure 4.2, three different types of carrier offices (edge, regional, and hub) are shown. These offices provide the end users with connectivity to either the Internet or a carrier's set of application servers. The carrier networks currently use their edge, regional, and hub central offices (COs) to provide telephone-switching services. As SDN is introduced into the network, these offices will start to transform into data centers. The type of equipment in these centers will start to change from specialty telephony equipment to generic servers, storage, and switching gear. Interconnecting all of these emerging data centers will be the carrier's optical transport networks.

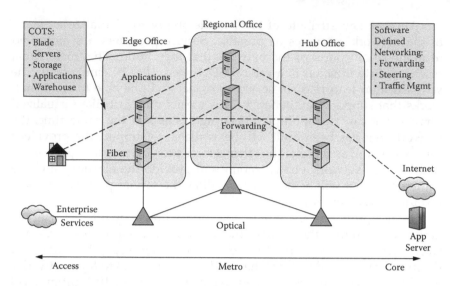

Figure 4.2 Deployment scenarios for carrier SDN.

Figure 4.2 shows various data flows. One such flow is the typical flow for a user to gain access to the Internet. This path will still traverse the network and will go through a gateway router. In addition, packet forwarding is shown. This forwarding is possible in software if the SDN functionality to support it has been implemented. Also, "service-aware" applications can be running and the routing of the packets for that service would be based on the type of service that is going to be provided.

This shows the vision that carriers have for introducing SDN into their networks. They are aware that this network migration will not happen immediately. Rather, pieces will be implemented over time until eventually the complete SDN has been put into place.

4.3   Telecommunication SDN services

Service providers do not view themselves as research and evaluation (R&E) labs where new networking ideas can be tried out and assessed from multiple perspectives. Instead, everything that a service provider does with its network has to be related to a business case and has to have a positive impact on the company's bottom line. The service providers only plan to build functionality into an SDN external controller if that functionality can be shown to have a significant benefit for the service provider.

Examples of this kind of functionality include new feature sets or new functionality that currently are not implemented into a protocol that the service providers are using in their existing legacy networks.

This can also include functions that could be performed using today's protocols but would end up having to be implemented inefficiently. Implementing a feature using SDN may also be done because it is believed that by doing so better scaling or economics can be realized. Finally, implementing a function in an external controller may be done if it can solve a problem that is currently not addressed by either today's network equipment vendors or existing network standards.

The traffic-steering function provides a good example of the type of function that service providers believe may be well suited to implantation in an SDN external controller. In this case, the service provider wants to be able to route the packets associated with a service based on the target service or application and not have to deal with doing any routing based on packet headers. The routing that is done for these packets may use the IP header information as one part of a larger data set that is used to make routing decisions. Other data that may be used could include real-time information, such as the network's current congestion status or subscriber profile information. Note that much of the information that will be used to make routing decisions for the packets will not be carried in the packet itself. The software routing will be responsible for pulling a great deal of information together and then finally making routing decisions based on that. This functionality could not be accomplished in the type of router that is used in today's legacy networks.

The use of the OpenFlow interface may assist service providers in implementing traffic steering in an external controller. An example of how this could work would be to look at the case in which a subscriber is watching a video over the network. Using the OpenFlow interface, the external controller could detect by inspecting the first few packets of the flow that a long-lived flow was being streamed over the network. Once this was detected, the underlying physical switches could be commanded via the OpenFlow interface to start to use cut-through switching to reduce the overall hardware needed to support the services that are delivered. By reducing the load on the system that is required to support this long-lived flow, the system is better able to provide a higher level of service to other users while the video flow is being transported. This gives service providers an efficient way to provide service-aware video services.

Hybrid cloud computing provides another example of a situation in which service providers believe that an SDN external controller could provide them with network functionality that is not available today. A hybrid cloud computing scenario occurs when an enterprise is operating a private cloud in its data center. When an enterprise determines that it has a need for additional computing resources for a limited period of time, it makes sense to temporarily expand its private cloud not by purchasing and installing additional servers, but rather by leasing existing cloud capacity from a service provider for a fixed amount of time.

One of the biggest challenges that enterprises face when they are looking for ways to connect their existing private cloud to a service provider's cloud is that a physical link between the two networks has to be established. What this means for the enterprise is that they are going to end up paying for bandwidth on a 24 x 7 basis that they will not be using most of the time. Depending on the company's specific needs, these links can be large (10 GB) and therefore expensive. This is one of the key reasons why service providers believe that the hybrid cloud computing service has not been adopted by customers as much as had been expected.

The SDN solution to this problem is to virtualize the network. Once the network is virtualized, then allocation of that network can be given to the control plane, which will then determine how much of the network resources the enterprise needs at any moment in time to connect its private cloud to the service provider's cloud. This then creates the possibility of offering bandwidth on demand as a service to enterprises that want to connect their private cloud to the service provider's cloud. The network would then determine how much bandwidth the customer needed at any point in time. This would allow the customer only to have to pay for what the customer would be using.

Service providers see the lower cost of the service as more appealing to more customers. The bandwidth to connect to the service provider's cloud now becomes a multitenant offering and turns into a time-sharing service. The service providers will no longer be making as much money from a single customer, but instead will be making more money from more customers.

Another situation in which service providers believe that SDN can provide them with network functionality that they do not currently have is in the case of adding OpenFlow functionality to existing legacy routers that are operating in their networks. Again, there will be no "flash cut" in which all of the existing routers will be removed from the network only to be replaced by SDN routers. This means that the service providers are looking for a graceful, incremental way to move from today's network to SDN in the future.

Adding OpenFlow functionality to a legacy router can be done several different ways. One way is the scenario in which a router already has its native control that has been provided by the manufacturer. To this configuration, OpenFlow can be added to create a hybrid mode by which both control sets exist within a single router. Each set of interfaces would operate separately; however, there would have to be a system established to allow the control sets to share the physical resources of the routers.

Finally, service providers are concerned about how SDNs are going to scale. Today's service providers operate large networks, and they realize that tomorrow's SDN environments will be just as large, if not larger.

This means that they need to find solutions to issues having to do with OpenFlow switch partitioning and the best way to support multiple SDN controllers.

Figure 4.3 shows an example of how traffic steering is being implemented in a service provider's network today. The traffic arrives in the network and is then routed from the traffic-shaping application to a series of applications. Initially, the packets will follow the application trail shown in Figure 4.3 as the network discovers the nature of the communication session. Once that is known, there may no longer be a need for all of the packets to be processed by all of the applications. After that happens, the remaining packets can follow the overlay trail and only be processed by the applications that truly need to see each and every packet.

The application "traffic-stitching points" exist in the traffic-shaping algorithm between the overlay trails. The purpose of this is to allow the development of arbitrary network feature graphs that can be allowed to vary over time. It is the purpose of the traffic-steering algorithm to determine the application trail that will be taken through both the service features and the cache.

This implementation of traffic steering does provide flexibility and can be considered extensible. However, it requires network packets to travel through a large number of network interfaces and requires a great deal of processing power to be spent processing each and every packet. This processing may not be required for every packet, and when long-lived flows are traveling over the network, this is a less-than-optimal solution

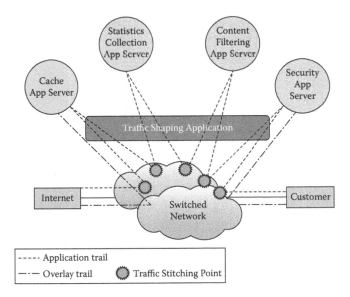

Figure 4.3 Traffic-steering methods.

that will use up too many network resources that could be better used processing other flows.

Figure 4.4 shows how a traffic-steering application might be implemented in SDN. One of the goals of this network design is to allow network packets to avoid being processed by the traffic-shaping application if that type of processing is not needed. In the example shown in Figure 4.4, the customer is retrieving data from the network's cache application server after the network has processed this data for security and content filtering. Once the network determines that it will be processing a long-lived flow, it can reduce the amount of processing that it does on each packet in the flow by starting to process the packets using cut-through routing.

Network statistics are still important, even for this type of flow that will not be processed by the traffic-shaping application, and the statistics for the long-lived flow will be collected by the OpenFlow interface. Another way of implementing this would be to have the OpenFlow controller determine flow usage for a particular pattern and use this information to determine when a flow completes.

In Figure 4.4, the OpenFlow controller is shown as a separate application. The OpenFlow controller application programming interface (API) could be implemented as a proprietary API. Alternatively, this software could be implemented as a part of the traffic-shaping application.

Figure 4.5 shows an example of a network that could be used to provide an enterprise with access to a service provider's hybrid cloud resources.

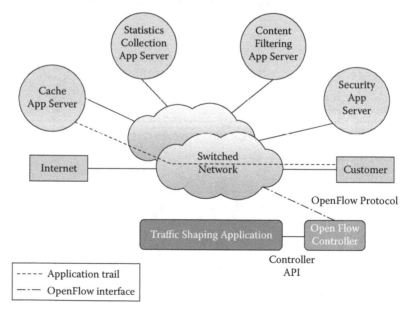

Figure 4.4 Using OpenFlow for traffic-steering optimization.

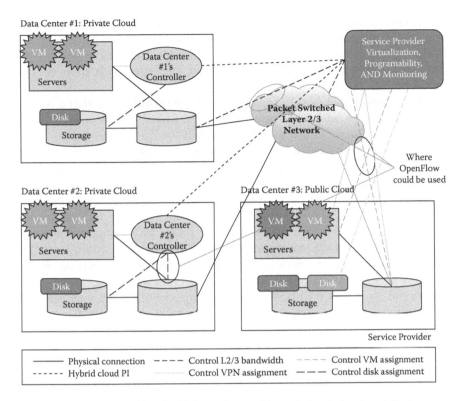

Data Center #1: Private Cloud

Figure 4.5 Example of bandwidth on demand for a hybrid cloud architecture.

In this scenario, OpenFlow could be used to organize how the data would be routed through the routers.

A key part of this figure is the hybrid cloud APIs, which do not currently exist. However, if the idea of a hybrid cloud is going to gain acceptance for enterprise customers, this part of the SDN technology is going to have to be developed by service providers and equipment providers to permit seamless integration between enterprise private clouds and service provider clouds. These APIs will allow multiple users to make use of the underlying network resources. This will determine how resources are requested and then controlled once they are granted.

Service providers are interested in using SDN to create the next generation of service provider networks. Their motivations are driven by the practical underlying economic realities of today's legacy networks, which are becoming too expensive to both build and maintain. The increasing cost of the network equipment being purchased and deployed in today's networks, coupled with the larger and larger volumes of equipment that the growing networks are requiring, is driving service providers to look for more economical ways to build their networks.

A network built using SDN technologies offers service providers the possibility of finally being able to get their network costs under control. Building a network based on a COTS infrastructure offers the service provider a way to align its network's cost structure with their service revenue.

Implementing SDN will fundamentally change the way that a service provider designs and deploys its networks. The COs that are a part of today's legacy service provider network will be transformed from holders of special-purpose telecommunications equipment into data centers. This will allow the service providers to be able to reap the benefits that come along with reduced cost, scaling, and new service flexibility that are offered by the arrival of the technologies associated with cloud computing.

The use of SDN technology to build the network will allow the network to become aware of the types of traffic traversing the network. It will be possible for some types of network traffic (e.g., video traffic) to be handled initially by SDN and then by cut-through switching. Fewer network resources will be used to transport these packages; therefore, additional resources will be available to support other network services and traffic.

The use of SDN technologies opens the door for fuller enterprise adoption of hybrid cloud computing. When implemented using SDN, hybrid cloud computing can use the OpenFlow interface along with new ways of managing network traffic to create a seamless interface between the enterprise private cloud and the service provider cloud. New hybrid cloud APIs will need to be created and adopted by the industry to fully enable this new service offering.

The greatest concern to service providers is the interaction of SDN technologies with existing legacy networks. Because of heavy investment in today's networks, service providers must find a clear path from current networks to the fully SDN-enabled environment of tomorrow. To make this happen, virtualization and resiliency changes will have to be made to the OpenFlow interface to ensure that SDN can coexist with today's legacy switching systems.

chapter five

SDN development

5.1 Introduction

The creation of networks based on software-defined networking (SDN) technology holds the promise of customizing, creating, and deploying networks that can be programmed and designed to ideally support a network's traffic. Although the possibility of implementing SDN technology to make networks more cost efficient is attractive, creating networks that can be tuned to the traffic that they are carrying is even more exciting.

5.2 Existing network limitations

The two major problems with today's networks are that the applications that are using the networks do not know enough about the underlying network that they are using and the network does not know enough about the applications that are using it. A wide variety of different techniques have been created to attempt to solve this problem; however, none of them does a good job.

Current approximation techniques are barely sufficient and are basically ineffective. Applications are forced to guess about the networks that they will be using to carry their packets. To make these guesses, applications have to rely on network tools such as ping-stats, Doppler, geolocation, and whois lookups. The network equipment that the applications may try to talk with all use different forms of proprietary codecs that require custom proprietary interfaces. The best that any application is ever going to do is to be able to create an approximate network topology and guestimate where their location within this network is. A much better way to go about getting the network information that an application needs would be for the application to be able to ask the network.

On the network side, existing networks have to resort to spying on the application traffic that they are transporting in the hopes of being able to gain a better understanding of the applications that are using the network. Networks have their own special set of tools that they use to do this: deep packet inspection, stateful flow analysis, application fingerprinting, and service-specific overlay technologies. All of this effort is being expended to allow the network to try to maintain the service-level agreements (SLAs) that have been promised to the applications that are using the network. This is expensive in terms of the network resources that it consumes.

One way to look at today's network is to view it as an interaction between four separate entities. These entities would be applications, application data, users, and the network. Each of these entities has a different view of what is going on:

- **Applications**: Applications are responsible for knowing what the capabilities of the end-user devices that have been attached to the network are because this is how the user will be interacting with the application. In addition, how close the end user is to the application's content is also critically important, especially for real-time applications such as games. Ultimately, the application will be responsible for controlling the network resources.
- **Content**: The idea that content can become network aware changes how the content will interact with the network. It will be the responsibility of the content to make modifications to its placement within the network along with controlling how content is selected. The actual insertion of content into network data flows can be controlled by network-based analytics.
- **User**: Perhaps the least complex of all four components. The end user is aware of what is wanted from the application and therefore from the network. To obtain the desired information, the application will be responsible for directing the user where to go on the network to get it.
- **Network**: The network lies at the center of the interaction between all four entities. This means that it is responsible for the real-time interaction between the user, the content, and the application.

To enable the interaction between these four entities, a new form of bidirectional communication needs to be implemented in the network. The applications have to be able to talk to the network, and the network has to be able to talk to the application.

5.3   Programmable networks

One of the most promising features of SDN technology is that the network can be programmed. The key functionality that a programmable network provides is the ability of the applications that are using the network to be able to inform the network about the desired, necessary network behavior. At the same time, the network needs to be able to inform the various applications that are using it about the data intrinsically contained within the network.

The way to permit this type of application/network communication to occur is to establish new network "touch points" (Ward, 2011). These programmable touch points will be programmed to create the network behavior that will be needed to support the applications.

The types of touch points that will be supported can be divided into two high-level groups. The first is user based, and the second is network based. User-based programmable touch points include such data sets as a user service profile, a billing profile, and a security gateway that could include both virtual private network (VPN) information and mobile information. The network-programmable touch points are focused on the behavior of the equipment that is being used to implement the network. These touch points include such items as the enterprise edge, business edge server profiles, content delivery networks (CDNs), and even hypervisor stack information to allow the management of how virtual machines (VMs) interact with the network.

5.4   Network/application information

In an SDN environment, applications will have access to information about the network that they are using that either is not currently available to them or is difficult to determine from the network. This new type of information will include information about the end-user devices that are attached to the network and are accessing the application.

Applications can be informed about where they are located in the network topology—the real location, what access network technology is being used, how much bandwidth is available, and the utilization of the link being used to connect to the end points. With these new types of network information, application balancing can be adjusted to better match the network's real-time usage. These types of adjustments may have an impact on how much the user is willing to pay for received services, and additional billing granularity must be supported. A final benefit of having greater access to more network information is that applications will then have more flexibility regarding where they are physically placed in the network.

Networks can become more efficient with the information that applications will be able to provide to them in SDN. Ultimately, applications will control the network resources. To do this, the network will have to use application-provided data to optimize both network bandwidth and network resources. This will allow new service topologies to be defined and supported. Security identification will be able to be included into the transport of all data across the network. Finally, with application-provided information, networks will finally be able to provide service-specific packet treatments.

5.5 Legacy to SDN

Today's legacy networks will not transform into SDN-based networks overnight. The evolution process that will occur is a topic of debate in the networking community. Mutual agreement is that it will occur by having the SDN technology augment what is already on the network. All participants agree that SDN must not break the networks that exist today.

As SDN-enabled equipment starts to be introduced into the network, there will have to be integration with the existing routing, signaling, and network policy logic. The true value of SDN is that it makes the network programmable. This means that it will need to support programmable touch points that are modular in design.

The service model that will be presented to users must be seamless. This can be accomplished by creating collaborative inputs. Ultimately, for the various service provider and vendor players to become involved, SDN products are going to have to be standards based.

Figure 5.1 shows how applications and the underlying network could communicate in SDN. The needed information will not be exchanged using a single protocol that has been loaded down with all of the application programming interface (API) information that is needed to support every application and every network. Rather, a collection of protocols can be used to provide the right application or the right network with the information that it needs. SDN will use a collection of protocols to provide a modular workflow approach to get information out of the network as well as to program information into the network.

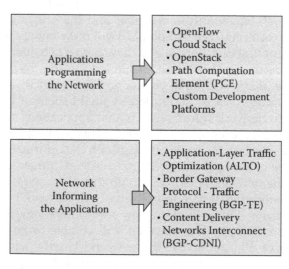

Figure 5.1 Applications and network interaction in SDN.

Providing this set of protocols will solve a number of the problems that exist in today's networks. Routers or switches in a legacy network do not currently have an API (or even a software development kit [SDK]) that allows them to program an access control list (ACL). This functionality currently has to be performed by a network administrator using a command line interface (CLI) or the Simple Network Management Protocol (SNMP).

Figure 5.2 shows one way that SDN could be made to work with existing legacy networks. In this example, the complicated network protocols and interfaces that are currently used are encapsulated in the network. On top of this, an orchestration layer is created that will provide an interface between today's collection of network protocols and higher-level web services-based applications, which would be the target interface for application software development. This abstraction allows the application developers to avoid the technical details of the networking-specific protocols.

Figure 5.3 shows how SDN and OpenFlow could be added to a legacy controller's control plane. In Figure 5.3, support for the OpenFlow protocol is being added to augment what is currently available in routers today. This figure shows the support for the OpenFlow protocol existing in parallel to the legacy protocols. This architecture has been referred to in various ways but is most commonly called the "ships-in-the-night" approach.

The ships-in-the-night approach to adding OpenFlow support to a legacy router would require that the router's physical resources be partitioned. A portion of these physical resources would be given to the OpenFlow functionality to control along with a portion of the router's ports. Another portion and a matching set of ports would then be given

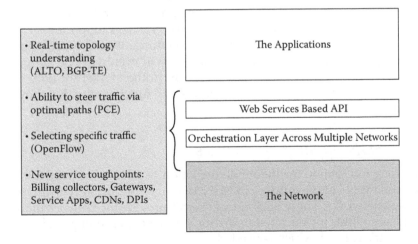

Figure 5.2 Proposal for the nondisruptive addition of SDN to legacy networks.

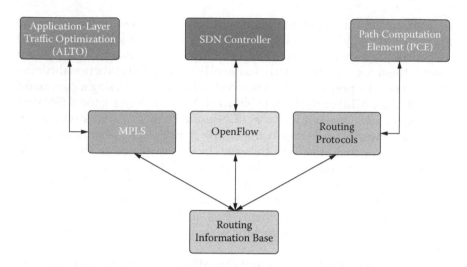

Figure 5.3 Adding SDN to a legacy router.

to the router's existing native control program to control as it does today. It is anticipated that there would be some level of integration between the OpenFlow and the native control program to ensure the proper operation of the router.

An alternative way of adding OpenFlow support to a router is to integrate the support for OpenFlow with the controller's existing legacy protocol logic. In this scenario, the OpenFlow functionality could be used to define features. This would allow the native control plane to be augmented with additional functionality. The integration of both types of control plane logic means that router resources would no longer have to be partitioned. This solution would allow the router to operate at several different abstraction levels depending on the type of network services supported.

The current belief in the networking community (Ward, 2011) is that both types of OpenFlow implementations will eventually be supported in SDN. The thinking is that there is room in the market for both types of solutions and the different price points and functionality that they will bring to the market.

5.6 SDN application services

A strong motivation for the implementation of SDN is the availability of network application services in SDN that are not currently available in today's networks. These SDN-only services provide the network management oversight, including support for service provisioning and routing and traffic control, which is needed.

5.6.1 Service-engineered path

Today's traffic-engineering paths are set up using protocols such as the Resource Reservation Protocol (RSVP). When program classifiers are requested using OpenFlow, this creates service-engineered paths, which permit the identification of the specific service requested from the flows. Once identified, the service request can be used in a service topology or to guarantee reserved bandwidth over that network, which can then be used to provide a new set of services.

A tunneling or switching technology is used to provide a path that is then used by specific functions of a given service. This will then allow the network to perform selective traffic redirection based on transient classifiers. These signaled paths would be set up using the path computational element (PCE) standardized API.

5.6.2 Service appliance pooling

For cost savings, network engineers can identify specific traffic and program the classifiers into the forwarding plane and take, on an average, 15–20 service appliances that are around an edge router today and pool them back into a data center. Once this is done, these service appliances can then be used to support multiple edge routers.

5.6.3 Content request routing

Content request routing has to do with identifying where the end user is located and then identifying where the data that the user is trying to access is located. The user-provided information will develop information based on the network infrastructure. This will need to be able to execute across multiple service provider networks, and it will have to support both mobile and broadband users.

The determination of which copy of the data should be used to support the end user will be based on a number of different factors. These will include network proximity, network availability, network congestion, the availability of the content, how much of a load the content will place on the network, and the capacity of the content. This type of information can be provided by the ALTO (Application-Layer Traffic Optimization) Protocol (Nadeau and Grey, 2013).

The goal is to allow the network to identify the best location to use for accessing the requested content. Within SDN, this answer can be provided by the network's topology; once the answer is known, it can be provided to the DNS (Domain Name System) server so that end-user requests are properly resolved.

5.6.4   Bandwidth calendaring

Within today's networks, there are preplanned activities that require a fixed amount of network bandwidth to support them. These types of activities can include adding flexibility to where services are placed within the network, the ability to schedule when data center content is backed up, management of the distribution of application content, and the orchestration of cloud-related activities.

Bandwidth calendaring refers to the ability to schedule the bandwidth that an activity is going to require. This means that a specific network path will be reserved to support an application. The use of SDN means that this path can be made available to the application when it is needed, and most important, it means the needed bandwidth will not be taken away when another high-priority application starts up in the middle of the application session. Only designated traffic will be permitted to travel over the reserved network path.

To implement bandwidth calendaring, four separate technologies will be required. The first is the use of the ALTO and BGP-TE (Border Gateway Protocol–Traffic Engineering) (Nadeau and Grey, 2013) protocols to provide a real-time understanding of the current network topology. Next, the PCE protocol (Nadeau and Grey, 2013) will be used to steer the application traffic through the optional network paths. The initial reservation will be established using a web services API. Finally, the specific traffic that will travel over the reserved path will be selected using the OpenFlow API.

5.6.5   Social networking

In the world of social networking, for a wide variety of reasons, where the end user is located is important. However, the technologies that are used in today's legacy networks can at best approximate the user's location. There is also a great deal of useful information currently unavailable to social media applications. This "missing information" includes the access technology that the user is using and the bandwidth that is currently available to them, the capabilities of both the device that they are currently using and the network connection through which they have connected, their specific geographical location, how close they are to network content, how much bandwidth they have paid for, and any security-related issues.

Today's networks attempt to determine where a user is located by using three different methods. The first is an active broadcast by the user of where they are located—effectively "checking in." The next is a so-called game broadcast that occurs when a user moves to a new access link. Finally, there is the passive derivation by which the network uses what information is available to try to determine the end user's location.

The goal in an SDN-enabled network is to allow social networking applications to provide continuous real-time streaming of their resources, the people that they are interacting with, their location, and their surrounding network content. The end user will be defined by data that includes their access, capacity, bandwidth, and their profile. In addition, because their location will be known, other information will be available, including content, resources, places, and people.

chapter six

Network vendors and SDN

6.1 Introduction

The arrival of the software-defined networking (SDN) approach to network design brings the potential to change how both carriers and enterprises build their networks in the future. This change could have a significant impact on the companies that provide the network hardware used to build the networks.

The network hardware used to build networks today is very sophisticated. Once a firm selects a specific vendor's equipment, the firm is committed to buying more equipment from that vendor as its network grows, along with the necessary software and hardware upgrades that occur over time. SDN may change all of this.

In SDN, simple high-speed routers are controlled by sophisticated software applications that run on separate servers. This architecture is completely different from what is deployed in networks today. To be able to survive and grow in this new networking environment, all of today's vendors are going to have to change significantly. This is a dynamic narrative that will evolve in the decades to come. The steps that each of the major equipment vendors are initially taking are summarized here.

6.2 Cisco

Today's legacy networks have been built using sophisticated routers that are provided by networking equipment companies. Cisco is one of the largest and most successful of these companies. Clearly, the arrival of SDN and its use of simple packet-forwarding hardware pose a significant threat to Cisco's primary source of revenue. Cisco's response to the threat posed by SDN will have a significant impact on how networking is done in both the short and long terms.

In November 2013, Cisco announced its initial response to SDN: the Application-Centric Infrastructure (ACI). In addition, Cisco realized the need for more expertise to create a strategy to deal with the threat posed by SDN, which resulted in Cisco's acquisition of the Insieme Networks company.

Cisco had already held an 85% ownership stake in Insieme and had made repeated investments in the company, including $100 million in April 2012 and an additional $35 million in November 2012. The end result of these investments was that in November 2013 Cisco announced it was going to fully acquire the SDN-focused firm.

The products that Insieme was developing were the Nexus 9000 line of data center and cloud switches that featured application awareness to make the network infrastructure flexible and agile and permit a dynamic response to application needs and virtual machine workload mobility.

The Nexus 9000 is Cisco's hardware-based response to SDN. The Nexus 9000 switches contain both "off-the-shelf" chip components ("merchant silicon") and custom silicon for both basic network virtualization (OpenFlow and VXLAN [Virtual Extensible Local-Area Network]) and Cisco proprietary application-centric networking, which is being called the ACI. The big difference between Cisco's approach and the open-source SDN approach is that Cisco wants the network to be application aware versus applications being network aware. ACI is hardware designed to equally provision both physical and virtual resources in data centers and cloud networks no matter what hardware or hypervisor is the basis for those resources.

The Cisco ACI fabric consists of the three components (Figure 6.1): the Cisco Application Policy Infrastructure Controller (APIC) and Cisco Nexus 9000 Series leaf and spine switches. In an ACI network, the leaf top-of-rack (TOR) switches attach to the spine switches and never to each other. The spine switches attach only to the leaf switches and possibly to a higher-level spine switch if the network design is hierarchical. The Cisco APIC (and potentially all other devices in the data center) attach to the leaf switches only.

In an ACI network, services are supported by providing a REST (REpresentational State Transfer) application programming interface (API)

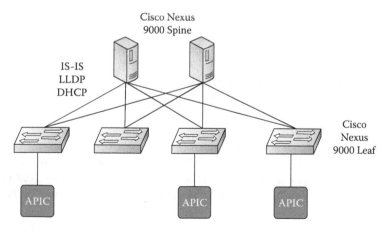

Figure 6.1 Cisco ACI fabric.

that can be used for automation. ACI provides a service graph that is composed of logical service functions that are communicating to the outside using end point groups. The ACI service graph characteristics include traffic filtering based on policy, taps, graph splitting and joining, traffic reclassification, and so on. ACI has the ability to provide resource pooling for stateless load distribution across multiple destinations, each of which can then perform its own stateful load balancing. Simple pooling is supported by devices that are unaware of ACI, and more advanced pooling can be provided by those that are aware.

Cisco's APIC provides ACI with the brains to implement both policy and management (Matsumoto, 2013). APIC is a critical part of Cisco's SDN plan. APIC does the following functions: assigning policy to a traffic flow, having the policy move with it, creating service chains to apply the policy properly, and automatically attaching policy to certain types of workloads.

The APIC is a cluster of controllers that have the following characteristics: They are distributed, they offer a single point of control and provide a central API, and they contain a repository of global and policy data. APIC policies are distributed under a variety of conditions. An example of this would be "just in time" when a node attaches or statically. Node attachment is detected by the APIC using triggers. Cisco has stated that the scalability of APIC includes 1 million plus end points, more than 200 K ports, and more than 64 K tenants (Matsumoto, 2013).

The Cisco Nexus 9000 router will also support the OpenFlow interface and the OpenDaylight open-source controller, but only in "off-the-shelf" chip component-based "stand-alone" mode on the Nexus 9000. Full-function ACI mode can only be achieved using Insieme's custom ASICs in the Nexus 9000. Not only does ACI require new switches (the Nexus 9000 line), but also the switches are not line-card compatible with the old switches (the Nexus 7000s), and they will require a software upgrade when ACI is fully implemented.

The OpenDaylight controller does not play any role in Cisco's design (Duffy, 2013). Insieme developed its own controller for ACI-mode networking, which is achieved by using Insieme's custom ASICs. In stand-alone mode, the Nexus 9000 with off-the-shelf silicon—in this case, Broadcom's Trident II—can support OpenDaylight, OpenFlow, and other open-source software for SDNs and network programmability and virtualization.

Even though Cisco's monolithic architectures may be criticized, it is likely that some customers will prefer them to other alternatives. Cisco is offering its customers a single vendor that provides all the pieces of the SDN solution, which is highly appealing to an industry that has been accustomed to single-vendor solutions.

Initial feedback from carrier customers regarding Cisco's SDN plans has not been favorable (Wilson, 2013). The US telecommunications provider AT&T released its Supplier Domain Program 2.0 Request for Information (RFI) at the end of 2013. In this document, AT&T asked its vendors, Cisco included, to identify if they were developing SDN controllers and how their network equipment will be adapted to be controlled within SDN. It has become clear that AT&T is interested in shifting away from purpose-built smarter hardware and toward commercial off-the-shelf boxes as part of its move to using SDN and network function virtualization (NFV) to save significant network capital expenditure (CAPEX) expenses.

Market research firms believe that AT&T is not going to be open to Cisco's ACI architecture, which includes the Nexus 9000 switches, because it still seems too complex and proprietary compared to architectures that are more white box oriented.

If AT&T decides to move completely away from ASIC-based hardware with some intelligence and toward a white-box approach that is being favored by software players such as VMWare, this will be an event of importance to Cisco.

6.3 VMware

VMware started the development of its SDN solution early with the $1.2 billion acquisition of startup Nicira in mid-2012. The motivation for this purchase was Nicira's network virtualization strategy, which fit well into VMware's overall product set, allowing for tight coupling with products such as vSphere. Just over a year after the Nicira acquisition, VMware announced its network virtualization platform, NSX, which had been created using Nicira technology in August 2013.

VMware, the leader in server virtualization, is using its NSX product to branch into network virtualization. NSX is a software-based network virtualization overlay that enables a VMware hypervisor to provide network control functionality as shown in Figure 6.2. It is the opposite of what Cisco proposes and should be viewed as a direct competitor to the Cisco/Insieme ACI and Nexus 9000 approach.

NSX provides network virtualization by provisioning hypervisor virtual switches to meet an application's connectivity and security needs. NSX uses virtual switches that are connected to each other across the physical network using an overlay network. NSX accomplishes this using a distributed virtual switch (the VMWare vSwitch product). The vSwitch sits at the network edge in the hypervisor and handles links between local virtual machines. The vSwitch provides access to the physical network if a connection to a remote resource is required (Banks, 2014).

The NSX controller arbitrates applications and the network. The NSX controller uses northbound APIs to talk to applications, which communicate their needs. The NSX controller then programs all of the vSwitches under NSX control in a southbound direction to meet those needs. The NSX controller can use the OpenFlow protocol for those southbound

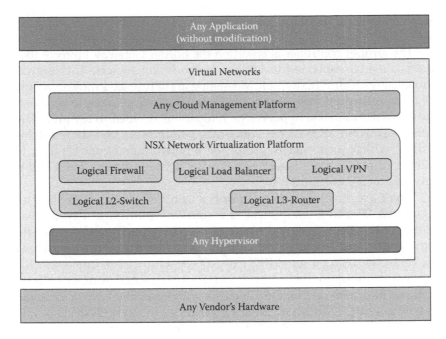

Figure 6.2 VMWare's NSX virtualization platform.

links, but OpenFlow is not the only part of the solution or even a key one. In fact, VMware deemphasizes OpenFlow in general (Banks, 2014).

VMware NSX creates a virtual network overlay that is loosely coupled to the physical network underneath. Cisco and other vendors argue this approach will not scale well. Network experts believe that whether software switches and virtual network overlays are enough to handle high-performance environments depends on the networking situation.

It is agreed that if a simple software switch is performing a fairly basic Layer 2 feature set, the vSwitches (software switches) in ESX can do a great job today. Other cases, such as a software switch performing more advanced services in an open-source hypervisor, will not be able to provide practically jitter-free, submillisecond performance. In this case, removing the hypervisor from the network path and replacing it with dedicated network hardware still makes sense (McGillicuddy, 2013).

6.4 Juniper

Juniper Networks purchased Contrail Systems, a startup maker of SDN software, in December 2012 for $176 million. Contrail's goal had been to create a controller that would be compatible with the OpenFlow protocols that had come out of Stanford University. However, the Contrail products were based on existing network protocols and would therefore be compatible with existing switches, routers, and server virtualization hypervisors.

The Juniper controller is based on the Border Gateway Protocol (BGP) that is already embedded in Juniper switches and routers. The controller also employs XMPP (Extensible Messaging and Presence Protocol), a protocol for transmitting message-oriented middleware messages, to control the virtual switches inside hypervisors. Juniper has decided to use an existing technology from telecom networks called Multiprotocol Label Switching (MPLS), which encapsulates packets on a network and controls their forwarding through those labels; MPLS exists between Layers 2 and 3 in the network stack (Morgan, 2013).

There are several parts to Juniper's Contrail platform, as shown in Figure 6.3. The first component is a software controller. This software controller has been designed to run on a virtual machine and supports a redundant active-active cluster configuration. Cloud tools can interact with the controller using a set of RESTful APIs that have been implemented to support northbound interaction. When it was launched, Juniper's Contrail platform had been certified to interwork with OpenStack and CloudStack, as well as with IBM's SmartCloud Orchestrator (Murray, 2013).

The second component of the Juniper controller solution is the vRouter, which can be run on either the KVM or Xen hypervisors. The Contrail controller will communicate with the vRouter using the XMPP, and this will permit it to tell the vRouters how to forward network packets. The vRouters

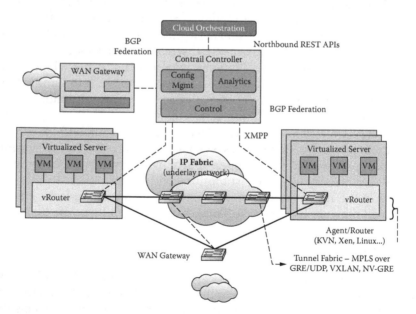

Figure 6.3 Juniper's Contrail controller.

will be responsible for building tunnels that run over the physical network between virtual machines that are active in the network.

The initial release of the Juniper controller will not provide support for the OpenFlow protocol. The plan is to adopt a wait-and-see attitude to determine how the market reacts to OpenFlow.

One of the biggest differences between an OpenFlow-based controller and Juniper's controller is that the Juniper solution keeps the master copy of the forwarding tables on the controller and copies them to the switches rather than keeping the master copies on the switches and aggregating them on the controller after they have been changed. Juniper, which is a member of the open-source controller project OpenDaylight, has announced that it will be making the code for its Contrail controller open source instead of contributing code to the OpenDaylight Project.

6.5   OpenDaylight

The OpenDaylight Project is an open-source project with a modular, pluggable, and flexible SDN controller platform at its core. This controller is implemented strictly in software and is contained within its own Java virtual machine (JVM). As such, it can be deployed on any hardware and operating system platform that supports Java.

The OpenDaylight Project is currently supported by 31 networking industry companies. These supporters include Cisco, IBM, Juniper, Microsoft, Redhat, VMWare Ciena, Intel, Dell, HP, and many more. The OpenDaylight community is developing an SDN architecture that supports a wide range of protocols and can rapidly evolve in the direction SDN goes, not based on any one vendor's purposes.

The OpenDaylight controller provides open northbound APIs that are designed to be used by applications. The OpenDaylight controller supports the Open Specifications Group Initiative (OSGi) framework and bidirectional REST for the northbound API.

The OSGi framework is used for applications that will run in the same address space as the controller; the REST (web based) API is used for applications that do not run in the same address space (or even perhaps on the same machine) as the controller. In the OpenDaylight controller, the business logic and algorithms reside in the applications. To perform their functions, these applications use the controller to gather network intelligence, run algorithms to perform analytics, and then use the controller to propagate the new rules to network devices, if any, throughout the network.

The OpenDaylight controller platform contains a collection of dynamically pluggable modules that perform needed network tasks. A series of base network services exists for tasks such as understanding what devices are contained within the network and the capabilities of each, statistics gathering, and so on. In addition, platform-oriented services and other extensions can also be inserted into the controller platform to provide additional SDN functionality.

The OpenDaylight controller's southbound interface has the ability to support multiple protocols. These protocols can be added as separate plug-ins. Examples of protocols that can be added include OpenFlow 1.0, OpenFlow 1.3, BGP-LS, and others. These plug-in modules are dynamically linked into a service abstraction layer (SAL) of the controller. The SAL interfaces device services to which the modules north of it are written. The SAL determines how to fulfill the requested service independent of the underlying protocol used between the controller and the network devices.

OpenDaylight's first release of their open-source controller, named Hydrogen, will include new and legacy protocols such as Open vSwitch Database Management Protocol (OVSDB), OpenFlow 1.3.0, BGP, and Path Computation Element Protocol (PCEP) (Nadeau and Grey, 2013; OpenDaylight, n.d.).

Hydrogen will also include multiple methods for network virtualization and two initial applications that leverage the features of Open Daylight: Affinity Metadata Service to aid in policy management and Defense4All for distributed denial of service (DDoS) attack protection.

To make OpenDaylight more cloud friendly, it will include an Open Stack Neutron, OpenStack's virtual networking plug-in, and the Open vSwitch Database project will allow management from within OpenStack.

6.6   Big Switch networks

Big Switch Networks is a well-funded network virtualization and SDN company that provides several products based on the OpenFlow protocol. Big Switch Networks has created three main products:

1. **Big Network Controller**: This controller is a commercial version of the open-source Floodlight controller. The controller is a platform on top of which software applications run. This automates the underlying fabric and allows an entire network to be controlled from a single console.
2. **Big Virtual Switch**: This component virtualizes the network using existing servers, enhancing network flexibility and making use of resources in the most efficient manner possible.
3. **Big Tap**: This is a network-monitoring application that uses OpenFlow-enabled switches to provide network administrators with full network visibility. This permits them to scale the network to minimize operating costs.

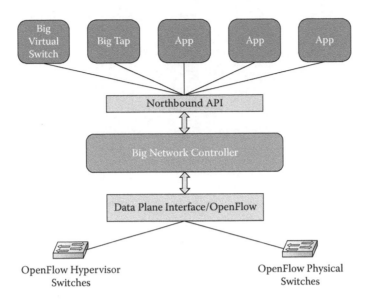

Figure 6.4 Big Switch Networks Open SDN.

Big Switch's Floodlight-based controller, shown in Figure 6.4, initially only talked to Open vSwitch switches and the unnamed virtual switch inside Hyper-V. It could support up to 250,000 new host connections per second and talk to over 1,000 physical or virtual switches on a single two-socket x86 server (Morgan, 2012).

chapter seven

Google and SDN

7.1   Introduction

Google is one of the most successful Internet companies of this century. Google's search engine technology and advertising products have provided immense financial success. The networks and the data centers used as a part of their business are critical to the company's success. With deep financial resources, Google can invest in any networking technology available or on the horizon to optimize network performance.

Therefore, it was a significant event in April 2012 when Urs Hölzle, Google's senior vice president of technical infrastructure and a Google Fellow, gave a presentation at the Open Networking Summit in Santa Clara, California (Hölzle, 2012). During his presentation, Hölzle stunned his audience by revealing that Google had built and was using a network based on software-defined networking (SDN) to interconnect its data centers. At the time of Google's initial implementation, there were no commercial vendors or products available for purchase. Google's engineers had created everything they used.

During his presentation, Hölzle revealed that Google's network contained two separate large-backbone networks. One of the backbone networks was designed to be Internet facing and is used to carry user traffic. The other backbone network was designed to carry internal Google traffic and is used to interconnect Google's data centers.

7.2   Earlier network management

Google's experience running its own network has reinforced the belief that running and managing networks is not an easy task—in fact, it is difficult. To simplify this complex task, OpenFlow has been implemented to accomplish two tasks. The first is to improve the performance of its backbone network, and the second is to reduce the complexity (and associated costs) of managing the backbone.

Google's networking needs are very large. Assessments estimate that Google is responsible for 25% of all Internet traffic in North America (McMillan, 2013). If Google were an Internet service provider (ISP), it would be the second largest ISP in the world. Many Google products, such as

YouTube, web search, Google+, Maps, and updates to the Android operating system and the Chrome browser generate a great deal of network traffic.

Hölzle stated that Google had run into a significant problem. In a data center environment, it is a general rule that as the scale of an application increases from a single server to multiple servers, the cost of running that application as measured in central processing units (CPUs) and storage decreases. However, the problem that Google encountered was that this same effect was not being realized when it came to networking. The cost/bit was actually increasing as the network usage increased.

Google revealed that wide-area network (WAN) unit costs have been decreasing as their WAN backbone networks have been growing (Hölzle, 2012). However, there has been a problem. Their WAN unit costs have not been decreasing fast enough to keep up with the surging increases in WAN bandwidth demand that they have been experiencing (Vahdat, 2012).

There were multiple reasons for WAN cost challenges. As the number of servers and storage devices that were involved in the network increased, the need to have the various network boxes talk to one or more other boxes at the same time increased. To accomplish this, networking equipment that was more sophisticated was required. As more and more networking gear was introduced into the network, the need to manually configure each of the separate boxes increased, so additional skilled technicians were required. Finally, because multiple vendors were supplying the various pieces of equipment used in the network, trying to automate the configuration of each of the pieces of network equipment using the various nonstandard vendor application programming interfaces (APIs) was turning into a significant task.

Experienced data center management companies, such as Google, have automation tools that allow automation of the management of thousands of servers. These powerful tools keep the management tasks rational, even as more and more servers are added to a data center. Google, like many other large firms, was searching for a way to replicate server management tools in the network management environment.

What was needed was a move beyond managing the individual boxes that were used to build the network to managing all of the boxes at once, as if the network were a type of "fabric" able to be managed as a single entity. The ultimate goal was to be able to manage the network based on the applications that were using it—to manage application traffic flows instead of network boxes.

This desire for a network fabric was thwarted by the equipment used to build today's networks. The protocols used in today's networks to control the forwarding of data packets are all box specific, with no concept of a network fabric. The network equipment that is being used has been optimized for transferring packets and not for monitoring the network or operations on the network itself. Finally, the network equipment in use is designed for the best case: Everything is working perfectly. Not enough attention has been paid to the cases for which the network is trying to deliver low-latency traffic or when there has been a network device failure, and a fast failover needs to occur.

7.3 Motivation for solution

To illustrate the network challenges, Figure 7.1 shows a sample network with three separate traffic flows (Hölzle, 2012). In this network, applications connected to three different routers (A, B, and D) all need to communicate with an application connected to router F. Their traffic patterns all take different paths, but they share a common network link as they travel from router E to router F.

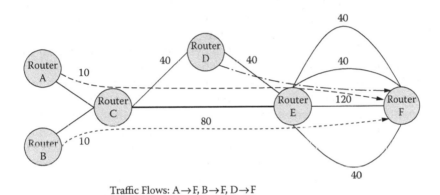

Traffic Flows: A→F, B→F, D→F

Figure 7.1 Sample network with three application traffic flows.

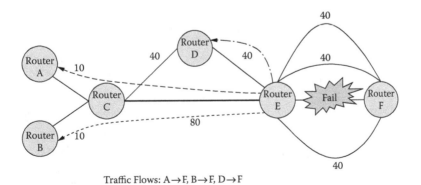

Traffic Flows: A→F, B→F, D→F

Figure 7.2 Network with link failure.

Figure 7.2 shows the same network experiencing a failure on the link that connects router E to router F. This failure will affect the traffic flows for all three of the applications that were using that link to connect to router F. In a modern network, when a failure like this occurs, each of the elements in the network will react to the failure. Router E will immediately send an informational message to all of the other routers in the network to inform them that the link between it and router F has failed. Each of the other routers will then automatically start to take actions to recalculate its flow tables. Each one of the routers responsible for sending packets from an application that is attached to it (A, B, and D) will now try to repair the path between it and the destination router, F.

Figure 7.3 shows how the network will look if router B is the first router that is able to secure a route to router F that uses the lower link between router E and router F. Note that the route that broke had enough capacity to simultaneously support all three traffic flows. However, now none of the remaining paths from router E to router F can support more than one traffic flow at a time. When router A and router D attempt to use the same path as router B, they will fail.

Finally, Figure 7.4 shows how the network might look after all three of the application traffic flows had been rebuilt. There are two significant drawbacks to the way that this type of recovery from a network failure is done in today's networks. The first drawback is that this process takes time. Routers will attempt to reserve paths, they will fail because other routers have already reserved the path, they will try to reserve another path, and they may either fail or succeed. It is not possible to predict how long it will take a router to rebuild a path to its end point when a network failure occurs. The networking term for the time that it takes to rebuild all of the impacted application traffic flows is *network settling time*.

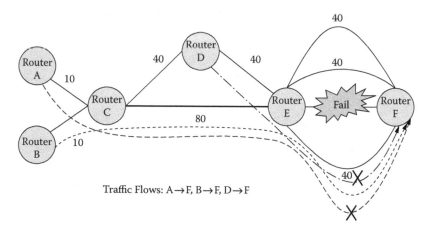

Traffic Flows: A→F, B→F, D→F

Figure 7.3 Network with one traffic flow repaired.

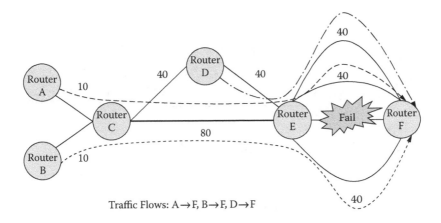

Traffic Flows: A→F, B→F, D→F

Figure 7.4 Network after all three traffic flows have been rebuilt.

The second drawback to how application traffic flows are rebuilt after a network failure is that the process is not deterministic. Depending on chance, any one of the three routers in our example could have reserved any one of the three remaining routes to router F. There is no way to accurately predict what the network will look like after a link failure. The problem with this approach is that sometimes the network may rebuild a set of links that fully support the needs of the applications that are trying to communicate, and sometimes it may not be able to do so. Each of the new links will have various characteristics that may or may not meet the needs of the applications that are using them. It is possible that, after a network failure, the network will not be able to create a new connection solution that meets the needs of all of the applications.

Network engineers at Google identified a better way to manage the link rebuilding process. Figure 7.5 shows the same network; however, this time it now contains additional component: a centralized traffic-engineering component. In this new scenario, the centralized traffic-engineering component is notified by router E when the network link between router E and router F goes down. The centralized traffic-engineering component is able to "see" the entire network. Using this information, it can now compute a new set of application traffic flows that will allow the needs of each of the applications to be met, and then each of the routers in the network can have its flow tables updated by the centralized traffic-engineering component.

Figure 7.6 shows how the network could look after having the application traffic flows rebuilt by the centralized traffic-engineering component. One important feature of this solution is that it is potentially much faster than the scenario in which each router autonomously builds its own new

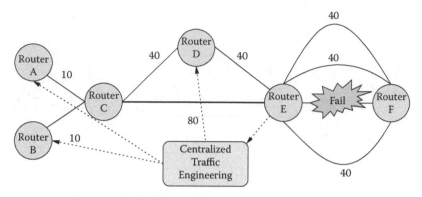

Traffic Flows: A→F, B→F, D→F

Figure 7.5 Using centralized traffic engineering to rebuild the network.

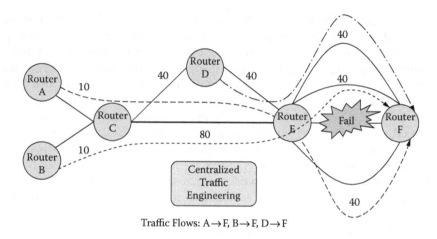

Traffic Flows: A→F, B→F, D→F

Figure 7.6 Rebuilt network with centralized traffic engineering.

routing solution. In addition, the network solution is now a deterministic solution—the same new network configuration will be produced each time assuming that the same inputs are provided to the centralized traffic-engineering component.

The idea of creating and using a centralized traffic-engineering component for Google's network was attractive for the following five key reasons (Hölzle, 2012):

1. Ability to create better network utilization scenarios if the design of the application traffic flows was created using a complete view of the current state of the network.

2. Decrease in the time that it would take to recover from a network failure because the centralized traffic-engineering component would have all of the required information and would therefore not have to retry solutions. This would greatly reduce the network settling time. The official term for this is *convergence,* and it refers to how long it takes the network to compute new routes that will meet the needs of all of the network applications after a network failure has been detected.

3. Increased control over how limited network resources were being used supports a decrease in the amount of overprovisioning needed because deterministic network behavior would be in place.

4. The deterministic behavior of the network would allow network engineers to create a separate "mirror" network that would allow more accurate simulation of the production network and test how the various event streams would both act and interact.

5. Overcome the computing power limitations of modern routers for processing new route computations. The centralized traffic-engineering component would have no limitations and is forecasted to be able to compute new network routes up to 50 times faster than a single router could. New configurations could be precomputed prior to occurrence of a link failure. Estimates are that new routes could be delivered to all routers in the network within 1 second plus the propagation delay between the central point and the routers.

7.4   Network testing

One of the biggest challenges that Google faced with its networks was that Google was unable to determine how any given configuration of the network would behave. Given that each router is a complex collection of hardware and software resources, the wide variability of inputs that such a network component could be receiving at any point in time meant that there was no way to predict how the network would react to a given set of inputs.

Google had been trying to create a test bed environment to try out network changes before they were introduced into the network. It was acknowledged that for any test bed network to provide realistic results, it would need to provide a full-scale replica of the production network. This was clearly impossible to do.

However, if a centralized traffic-engineering component was used to manage how the network's router traffic flows were calculated, then the current configuration of the traffic-engineering component could be duplicated in a test environment. This would allow real production inputs to be used by the network engineers to both research new ideas and at the same time try out new network configuration plans.

The use of the centralized traffic-engineering component has provided network test engineers with four advantages that were not available in traditionally designed networks:

1. Testing of network changes can be done in an isolated fashion. The availability of various logical models lets network engineers effectively do "network unit testing" with changes before determining what their impact on the entire network will be.
2. The centralized traffic-engineering component has a complete end-to-end view of the network, and this allows network engineers to test how their network changes are going to have an impact on the entire network. In this type of simulation, everything involved is real with the exception of the actual network hardware.
3. This type of lab testing allows network engineers to engage in performing built-in network consistency testing. They can cause two network links to fail at the same time to determine how the network will react. They can cause a network element not to fail but to start to misbehave and then determine what its impact on the network will be. The use of real-world network data allows this consistency testing to be done both in the test lab and in production.
4. The configuration of each device in the network is critical to the correct operation of the network. The ability to test the correct operation of the network has allowed the development of various networking tools that permit the state of each router in the network to be tested. In the lab, each router can have its traffic flow table validated after it is updated by the centralized traffic-engineering component.

7.5   Simulating the Google WAN

By choosing to implement a centralized traffic-engineering component, Google has greatly simplified the task of testing all of their planned network changes. Because all of the router flow tables are calculated in a single location, the Google network engineers are now able to simulate a portion or all of Google's WAN that interconnects the Google data centers.

Google is now able to treat how its WAN is configured in a similar fashion to how it manages large distributed software projects. To ensure that Google is able to accurately predict how the production network will react to planned changes, real production binaries are used in the simulation environment. The actual code that runs the production centralized traffic-engineering component is used. In addition, the actual binary implementations of the OpenFlow interface that will be running on the routers in the network are used.

To be able to simulate a real-world Google network, a large number of production servers have to be replicated in the testing environment. To do this, all of the routers that are being used in the network are simulated in the Google test environment. The real OpenFlow binary code is being used to make sure that the simulated switches behave like the real ones will; however, the router's hardware abstraction layer (HAL) is fully simulated. This introduces the limitation that performance of the real-world network cannot be accurately measured in the simulation testing laboratory environment.

One of the most powerful features of Google's simulated WAN testing environment is that the Google engineers are able to simulate any arbitrary topology. This includes building an accurate model of the entire Google inter–data center backbone network as it currently exists and simulating that configuration. In this simulated network, captured event streams from the real production network can be fed in to simulate actual events that happened in the production network.

An important benefit that Google has realized from creating such an accurate simulation environment is that Google is now able to test monitoring software in its simulation environment. Because production code is running in the simulated environment, the monitoring software can be plugged in to the WAN testing environment software and it will produce the same types of results that Google would see in its production environment. The use of the production-monitoring software allows the alerting conditions and monitoring software also to be tested in the simulated environment.

7.6   Google and SDN

Enterprise networks are a critical part of Google's ability to deliver services to customers and therefore they are an important technology that may provide a competitive benefit. Google's engineers investigated the current state of network technologies. They reached the conclusion that the arrival of SDN offers the greatest return on investment in the near term. Specifically, they have identified four key benefits that they believe that implementing SDN in their networks can provide to them.

The first benefit is that the implementation of SDN will allow the separation of network hardware from network software. No longer will both products be required to be purchased from the same vendor because one will not work without the other. Instead, network hardware can be selected exclusively based on the necessary hardware features. Unnecessary hardware features will not need to be purchased. Likewise, network software can now be selected based on the protocol features that will be used in the network and not based on the hardware that only supports that type of software.

The use of a centralized traffic-engineering component will allow logically centralized network control on a server that can be 25–50 times faster than the processing capability found inside a standard router. This will provide three critical benefits to the performance of the network. When network failures occur, the network will rebuild its links in a more deterministic time. The rebuilding of links will not suffer from the link congestion issues that require multiple attempts to create a new path between source and destination and therefore will be more efficient. Because the centralized traffic engineer component can have a hot standby, the overall solution will be more fault tolerant.

In today's network, network-monitoring functions are provided by the routers themselves. This means that, in addition to the packet-forwarding and flow table calculations that the router takes care of, it also has to support management functions. In Google's SDN implementation, this functionality is separated from the router functionality, and management, monitoring, and operations functions can be provided independently of the routers themselves.

The arrival of SDN technology brings with it a great deal of promise because the centralized network data and the network architecture allow it to be programmed now. Standard software development methodologies can now be applied to creating enterprise networks. This heralds a new era of both networking innovation coupled with enhanced flexibility that promises to change how networks are built and managed in the future.

By implementing SDN technology at the heart of mission-critical enterprise networks, companies will be able to achieve multiple benefits. A WAN will be created that is not only able to provide higher performance but also at the same time will be more tolerant of network faults and should turn out to be cheaper to both build and maintain.

7.7   Google's G-Scale network

When considering how to implement SDN, Google began with a review of its enterprise networks. Google operates two separate backbone networks as a part of its business (Hölzle, 2012). The first network is referred to as "I-Scale," and it is the backbone network that is Internet facing and carries user traffic for services such as Google search, YouTube, Gmail, and so on. This network looks similar to any ISP network and has been constructed using a standard array of commercially available networking gear from the leading network equipment vendors. The I-Scale network has demanding availability and loss sensitivity requirements that it must meet. The other backbone network that Google operates is called the "G-Scale" backbone network.

The G-Scale network is used to interconnect Google's data centers. Although it also has demanding network requirements because it is used for tasks like server backup and search index transport, there is more flexibility in how it performs. The traffic on this network is much more bursty than the traffic carried by the I-Scale network (Vahdat, 2012). The network consists of worldwide data centers that are interconnected over 10G links. This was the network selected for the initial implementation of SDN (Hölzle, 2012).

7.7.1 Google's G-Scale network hardware

The router hardware that has been used by Google to build the G-Scale backbone network comes from custom-designed configurations. When the networking project was started in 2009–2010, it was not possible to purchase a router that supported the OpenFlow protocol (Dix, 2012). The most important thing to realize about these routers is that despite their use to build a critical enterprise network, there is nothing unique about them. No custom application-specific integrated circuit (ASIC) chips have been used.

The G-Scale routers have been built using off-the-shelf silicon. Each of the routers is able to provide the network with 128 ports of nonblocking 10GE connectivity. Because the routers are used as part of an SDN environment, each router supports an OpenFlow networking protocol agent. What makes these routers unique is that outside the software needed to boot the box and the OpenFlow Agent (OFA), they run little other software.

The routers do not even support a command line interface (CLI). Everything, including the CLI, runs on the central control server, which is a multicore server with a great deal of memory. In addition, the central controller contains open-source software stacks that permit the support of the Quagga network routing software suite. Quagga allows the router to support both the Border Gateway Protocol (BGP) and the Intermediate System-to-Intermediate System (ISIS) protocol. When BGP runs between two peers in the same autonomous system, it is referred to as the Internal BGP (iBGP or Interior Border Gateway Protocol). The ISIS/iBGP protocols are used for internal connectivity. However, because Google could specify what functionality the routers have, the central controllers do not support unneeded networking protocols such as AppleTalk and Multiprotocol Label Switching (MPLS; which was supported on Google's legacy G-Scale network).

To provide the network bandwidth that the G-Scale backbone needs, each Google data center contains multiple chassis. This allows fault tolerance to be built into the network by having chassis provide backup for other chassis. In addition, support for multiple chassis per site allows the G-Scale backbone to scale to multiple terabytes of networking capacity.

Figure 7.7 shows how Google's data centers were interconnected using the G-Scale backbone network. In each data center, there are multiple Google

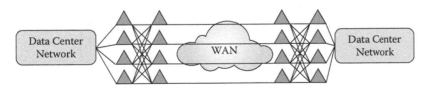

Figure 7.7 Google's G-Scale backbone network WAN deployment.

OpenFlow switches that are cross connected with each other. A set of long-haul optical transmission links then connects routers in one data center to the routers in the other data center. The routers are custom hardware configurations that are all running the open-source Linux operating system.

7.7.2　Google SDN deployment

A difficult, but not unique, challenge in SDN deployment was how to introduce the SDN technology into Google's G-Scale backbone network. The functionality that the G-Scale network provides to Google is not customer facing; however, it is critical to the 24 x 7 operation of the company. The network could not be offline or unavailable to retrofit it with the SDN technology.

In spring 2010, Google started phase 1 of its rollout of the Google SDN solution (Hölzle, 2012). The first step was to introduce the new OpenFlow-controlled switches into the network in parallel with their existing network hardware in three of their data centers. The plan was to make the new switches look like the existing routers that had already been deployed in the network.

This meant that there was no change in operations from the perspective of the G-Scale's existing non-OpenFlow switches. The BGP, ISIS, and Open Shortest Path First (OSPF) protocols were used to interface to the centralized OpenFlow controller (OFC) to allow the state of each of the controlled OpenFlow switches to be programmed.

Figure 7.8 shows what a typical data center network component may have looked like prior to the introduction of the SDN components. In this depiction of the data center network, at the edge of the network there exists a cluster border router. This border router communicates with the

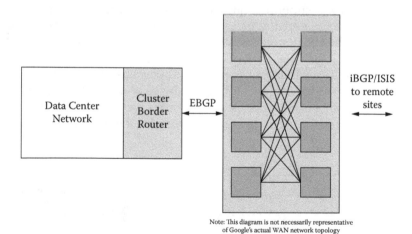

Note: This diagram is not necessarily representative
of Google's actual WAN network topology

Figure 7.8 Mixed SDN deployment.

Google core network using the Edge Border Gateway Protocol (EBGP). This is the protocol that is used to provide connectivity between data centers. The cluster border router is peered with a group of switches at the edge. The switches then connect across the WAN to switches at other data center sites. The iBGP and ISIS are used for internal communication between different data centers.

Figure 7.9 shows the same network; however, now a set of SDN control applications has been added. These applications were added on a separate server to move time-critical network protocol calculations off the processors that are embedded in the network routers and provide them with their own high-speed computation platform. This set of control applications includes the follow applications (Vahdat, 2012; Jain et al., 2013; Chandra, Griesemer, and Redstone, 2007):

- **Quagga**: Quagga is a suite of network routing applications that provide implementations of OSPF, Routing Information Protocol (RIP), BGP, and ISIS for Unix-like platforms.
- **OFC**: The OFC manages the OpenFlow protocol. OpenFlow is a communications protocol that is used to give access to the forwarding plane of a network switch or router over the network.
- **Glue**: Glue allows the OFC to talk with the Quagga application.
- **Paxos**: Paxos is a family of protocols for solving consensus in a network of unreliable processors. Google uses Paxos for leader election. The Paxos family of protocols includes a spectrum of trade-offs between the number of processors, the number of message delays before learning the agreed value, the activity level of individual participants,

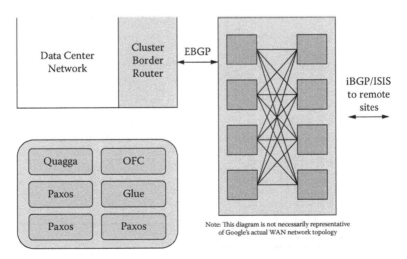

Figure 7.9 Mixed SDN deployment with control applications.

the number of messages sent, and types of failures. Paxos is usually used where durability is required (e.g., to replicate a file or a database), in which the amount of durable state could be large.

The next step in the process is shown in Figure 7.10; the protocols that the new application server will use to communicate with the rest of the network are added. This allows OFAs to begin to run on a subset of the switches that are controlling the network. The OFC can be connected to the OFAs using a variety of methods, including upgrading half the network at one time, doing every other switch, or just doing a pilot set of switches initially.

Once this change was made, Google had effectively divided its backbone network into two parts. The first part was the traditional backbone network. In this network, the network uses the EBGP to talk to the cluster border router. ISIS can be used to communicate with other WAN sites. No OpenFlow protocol is used in this configuration.

The other half of the network that consists of the new OpenFlow nodes is now speaking EBGP through Quagga on the high-speed server. In addition, iBGP/ISIS is used through Quagga on the high-speed server.

Forwarding table entries on the individual switches in the network are being coordinated through the OFAs. Quagga is able to do this because of the information that it receives from the OFC application via the Glue application that is used to connect Quagga and the OFC.

The process that network engineers used to deploy the new OpenFlow switches was performed on a site-by-site basis. The entire backbone network did not have to be upgraded simultaneously, and individual sites

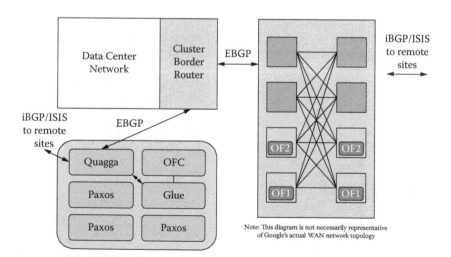

Figure 7.10 Adding protocols to allow the server to communicate with network.

within the network did not have to be fully upgraded all at once. The key to this upgrade technique was that the applications that were running in Google's data centers would not detect any differences between the OpenFlow routers and the non-OpenFlow routers.

The first step was to predeploy the equipment at a given site. Next, 50% of the site's bandwidth would be taken down. The engineers would perform the needed upgrades, after which the bandwidth would be brought up using the new OpenFlow switches. As a final step, testing of the new functionality would then be performed. Some of the network's optical data center interconnections were then moved over and made available to the new OpenFlow routers. As the new SDN functionality was introduced, the sites that had been upgraded to SDN provided full interoperability with the legacy router sites. This process was then repeated at a total of three of the G-Scale network's sites.

To test the new network that had been created, network engineers carefully selected applications that they would allow to make use of the OpenFlow switches and the associated network bandwidth. Low-risk applications, such as a copy service, were permitted to opt in and use the new network.

Once the entire backbone network had been upgraded to use the new SDN technology, the next step was to duplicate the functionality of the legacy network (Figure 7.11). The first task in this process was to add a traffic-engineering server that interfaced to the OFC module of each of the SDN routers. This traffic-engineering server was then able to have a global view of the communication patterns that were being used in the backbone network.

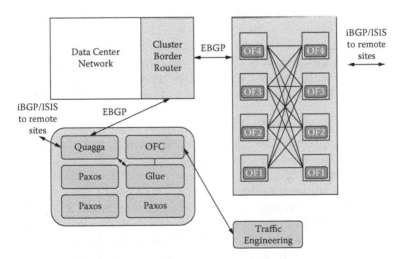

Figure 7.11 Adding additional functionality to an SDN environment.

Phase 2 of the deployment of the OpenFlow technology in the G-Scale network took roughly another year. During this time, the network engineers proceeded to activate a simple SDN environment with no traffic engineering. The network was extended to all of the G-Scale's sites; however, the SDN environment was still being operated in parallel with the traditional G-Scale network.

To continue to test the performance of the new network, more and more internal traffic continued to be moved over to the OpenFlow switches. This real-world network provided the Google engineers with opportunities to test the ability to transparently introduce centralized controller software updates without having an impact on network traffic. In addition, failover of the centralized controller was tested.

Phase 3 of the project started in early 2012. As part of this phase, SDN became part of the full-production G-Scale network at one site. Once this was successfully done, the rollout continued to additional sites. All of Google's data center backbone traffic plus a portion of the Internet-facing traffic that was found to be suitable for this network is now carried by the new OpenFlow network. The old network that had been built with legacy switches has been turned off.

The centralized controller has been upgraded via a major software release that includes support for the traffic-engineering functionality. This means that when the controller is establishing network paths, it can take into account the bandwidth needs of the various applications involved and ensure that each application will get the bandwidth through the network that it needs. Routing is optimized based on seven application-level priorities. This allows the centralized controller to provide globally optimized placement of network flows.

An example of the new types of functionality that Google has been able to implement because of their use of the SDN technology in the G-Scale network is support for bandwidth reservations. The Google application that schedules the creation and transport of copies of large data sets is able to interact with the central OFC. This interface allows the copy scheduler to work with the OFC to implement deadline scheduling for large data copies.

7.7.2.1 Bandwidth brokering and traffic engineering

The implementation of the SDN backbone network technology changed the way that Google was able to manage the bandwidth used in its G-Scale backbone network. Specifically, the new technology now allowed both brokering and engineering bandwidth use on a global basis.

In a typical data center, at any point in time there are a large number of applications that are running simultaneously. Each one of these applications would like a portion of the backbone network's bandwidth to be allocated to it so that the application can use the bandwidth to communicate with other applications. Figure 7.12 shows an example of the types of

	App 1	App 2	App3	App 4	App 5	App 6	App 7	App 8	App 9	App 10
App 1	–	27	–	–	–	65	–	71	24	18
App 2	78	–	22	71	35	71	53	66	12	–
App 3	–	–	–	82	50	–	–	–	–	8
App 4	24	92	16	–	83	27	26	56	–	–
App 5	5	51	60	84	–	9	12	20	42	79
App 6	93	13	17	–	78	–	20	90	3	–
App 7	–	57	–	51	–	87	–	69	–	–
App 8	39	21	80	63	22	34	43	–	66	10
App 9	38	30	22	56	8	79	51	26	–	61
App 10	–	–	–	–	47	–	–	–	–	–

Figure 7.12 Network bandwidth requests for communication between applications (Mbps).

bandwidth requests that could exist between a subset of the applications running in the network. Ultimately, there is more demand for network bandwidth than there is bandwidth available to use. In addition, not all of these applications have the same level of importance. Some applications, such as backups, need to be done but do not necessarily need to be done right now.

This means that the limited amount of G-Scale network bandwidth needs to be brokered among the various applications that are requesting it. The high-priority applications will receive more bandwidth to use, and the lower-priority applications will receive either no bandwidth or less bandwidth.

The network bandwidth will be allocated among the various applications using Google business rules by the Traffic Engineering and Bandwidth Allocation Server shown in Figure 7.13. The Bandwidth Requests Collection and Enforcement portion of this server is responsible for collecting the G-Scale network bandwidth requests from all of the various applications that want to use the network. This module has the ability to implement rate limiters on applications—if they have exceeded the amount of bandwidth that the Google business rules had allocated to them, then the amount of bandwidth that the application will receive in the future could be reduced.

All of the bandwidth requests are then forwarded to the traffic-engineering server module. It is the job of this component to create a solution for how application bandwidth should be allocated in the network. The bandwidth allocation results of the traffic-engineering server will then be communicated via an SDN API to a gateway to the SDN sites. This information will then be used to program the forwarding tables in each of the SDN routers in the network.

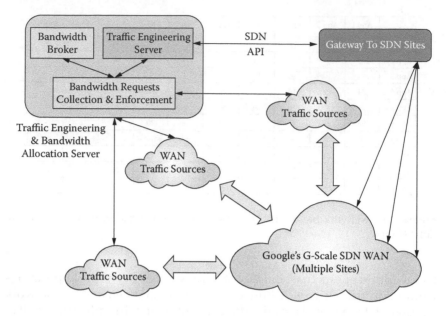

Figure 7.13 Google's high-level SDN architecture.

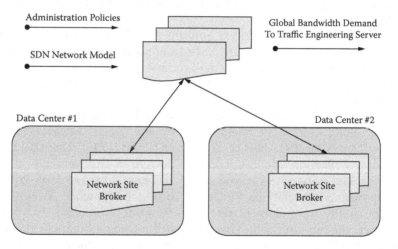

Figure 7.14 Google's SDN high-level architecture.

Figure 7.14 shows the architecture of the bandwidth broker component that is being used in the Google network. This is then used to communicate global demand to the network's traffic-engineering server application.

Figure 7.15 shows the service architecture of the traffic-engineering server used in the G-Scale network. The network's global bandwidth broker module starts the traffic-engineering process by passing it a "bandwidth

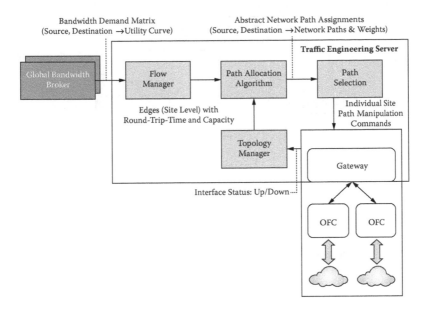

Figure 7.15 Traffic engineering server's service architecture.

demand matrix," which consists of source and destination applications along with their requests for bandwidth. Note that this is similar to what was shown in Figure 7.12.

In a legacy network, the path that would be constructed to connect two applications that wanted to talk with each other would be based on the shortest-path-first algorithm. In Google's SDN implementation, the path allocation algorithm is able to determine, based on a non-shortest-path-first basis what the optimum route through the network would be to connect two applications that want to talk to each other.

This route will be calculated by the traffic-engineering module based on its real-time knowledge of the network's topology. Any known network failures will have already flowed up through the network and will already be incorporated into the path allocation algorithm.

The end result of the traffic-engineering module will be a set of path assignments. This will indicate which source should talk to which destination using which path. This will be the information that is then sent to the gateway at the SDN sites that was shown in Figure 7.13.

Figure 7.16 shows what is happening in a specific Google data center. Each one of the switches is executing an OFA application. In this data center, there is an OFC, which is charged with managing multiple SDN switches (Vahdat, 2012).

The controller is receiving inputs from two different types of sources. The first of these source types has to do with routing: BGP and

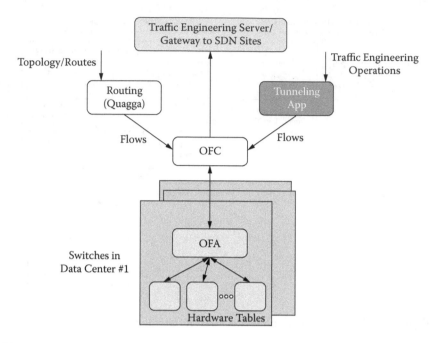

Figure 7.16 Architecture of an SDN controller.

ISIS protocol information. These sources will be providing the controller with the default shortest path information. This is basically the routing information that it would have if there was no traffic-engineering functionality in the network. The second source is a tunneling application that will be providing traffic-engineering operations information to the controller.

In Figure 7.17, a sample network site is shown with three SDN routers. An application that is connected to router 1 wants to connect to an application that is connected to router 3. Using the information that has been provided via the routing information source, router 1 could be programmed by the site's controller to create a direct path from router 1 to router 3—this would correspond with the shortest-path-first solution. The applications that would have their information sent over this path would be the ones that were either high priority or had connections that were latency sensitive.

The information that has been provided via the traffic-engineering system and the tunneling application can then be used to set up other network routes from router 1 to router 3 that pass through other routers in the network. Note that these routes will not be shortest-path-first routes. The applications whose information would be sent over this path may not be as high priority or may not be as latency sensitive.

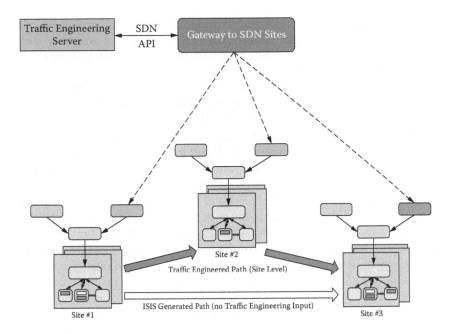

Figure 7.17 Use of an SDN controller in Google's backbone network.

7.7.2.2 Deployment results

One of the important discoveries that Google made as a result of introducing the SDN functionality into its network was that they were able to achieve a much faster introduction of new network functionality or bug fixes (Hölzle, 2012; Neagle, 2012). Google's production-grade traffic-engineering functionality was added to the SDN environment in just 2 months.

There are two reasons for this increase in speed. First, there were fewer devices that needed to be updated. Instead of hundreds of routers that are currently processing network data and have to be updated on the fly, now only a few control servers have to be updated.

Updating applications is a key part of how Google runs its main business, so there are many existing tools that Google can use to perform this function. Hitless software upgrades and new feature additions to the G-Scale OpenFlow network are able to be performed. These changes do not cause any packet loss or any network capacity degradation while they are being performed, primarily because most of the feature releases do not have an impact on the actual network switches themselves.

In addition, the Google simulation environment allows the entire network to be simulated. This functionality allows the network engineers to do much more extensive testing prior to rolling out the new functionality. One discovery that they made is that if a solution for placing all of the required network flows exists, the centralized network controller will find it. In a traditional network, a new network flow solution might theoretically exist, but because of how the protocols worked independently, the correct solution would never be found.

Google has stated that their ultimate goal is what they refer to as "push on green" (Hölzle, 2012). This means that when they have a software build that passes all of its submit tests, they should be able to instantly push it out into the production environment without any delay.

Security is an important part of operating any network. Google believes that SDN is more secure than a traditional network. The reason for this is because in SDN, it is only the central controller that needs to be secured. This controller can be isolated from the rest of the network and can be restricted to only talking with the switches in the network. These communication sessions can be secured through the use of certificates. In a traditional network, if any one of the routers was compromised because someone was able to log in to a switch, then the entire network would be at risk because additional routes could be inserted into its flow table. This is not a problem in SDN in part because no one is permitted to log in to the network switches.

Google reported that as a result of implementing SDN, it is already seeing higher utilization of the G-Scale network. Google has seen network utilizations of the links that connect its data centers that were close to 100% for extended periods of time (days). This is to be compared to traditional legacy WAN backbone networks, which normally experience link utilizations of 30–40% (Vahdat, 2012).

This high degree of network utilization is occurring because the new SDN technology allows for flexible management of end-to-end paths for maintenance. Google believes a traditional network could be "tuned" to provide the same type of deterministic network-planning results that they are able to achieve with SDN. However, the amount of time that this would require makes it impractical to do in the real world.

An additional benefit of the use of the SDN technology is that more application data can be exchanged via the network and less operations data has to be exchanged. Google has discovered that, in the SDN backbone network, there is a significant decrease in the amount of protocol-related traffic that must be transported. When the volume of tunneled traffic-engineering changes is compared to the traditional amount of ISIS protocol traffic that the network has to carry, Google was able to report a sixfold decrease in the amount of traffic that was transported by the network (Vahdat, 2012). This savings is attributed to the fact that there no longer is a need to operate at the link level of network granularity. The stability of the network is very high. The service-level agreement (SLA) for the internal backbone network is being met by the new SDN environment.

Moving forward, there are several different immediate opportunities that the new SDN technology is going to enable. With SDN technology it will be much easier to understand what is going on in the network. A unified view of the network fabric will be provided. If network engineers want to know what would happen under a given set of circumstances, all they have to do is go to a simulated network-testing environment and receive an answer to any questions. This insight is aided by the enhanced predictability that the network now provides. A network engineer can now take a look at the flow tables and understand how a given flow was created.

The traffic-engineering functionality that Google has been able to implement with SDN has yielded a number of real-world benefits. The first of these is that network engineers now have higher awareness of the quality-of-service (QoS) that the network is providing along with enhanced predictability of failures.

Figure 7.18 shows a sample three-node network that is experiencing a network failure of node 2. In a legacy network that did not have the traffic-engineering functionality that the SDN can support, the link failure could take up to 9 seconds to resolve. Both the detection of the failure and the convergence to a new network configuration solution would both take longer than in SDN.

Google believes that the delay within the traffic-engineering server in an SDN environment would be less than the ISIS protocol timers that would be used to detect and then communicate the failure. There are workarounds to these delays, such as "fast failover," but the network path solutions that they will create are not guaranteed to be accurate or optimal. It is Google's experience that in SDN this type of network failure can be resolved with an optimal new network configuration within 1 second (Vahdat, 2012).

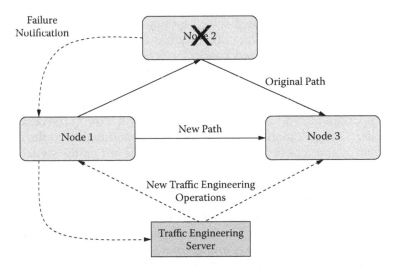

Figure 7.18 WAN convergence under failure conditions.

SDN is providing enhanced packet latency, packet loss, bandwidth utilization, and deadline sensitivity measurements. In addition, all network applications are no longer treated equally. SDN allows differentiation between applications and their unique network needs, resulting in improved routing between switches for applications. This can be provided because the SDN's central controller has a priori knowledge of the network's topology and the L1/L3 connectivity that exists in the network.

Finally, software tools have been created to provide improved network monitoring and alerts. The use of a centralized controller allows many of the monitoring tools to be connected to the high-speed server, which is able to provide network routing data in real time.

7.8   Implementation challenges

Although Google was able to successfully implement and operate SDN, a number of challenges with this new technology have been discovered as a part of the project. One of the most significant issues encountered was that the OpenFlow protocol is still in its infancy (Hölzle, 2012). Not all of the desired functionality has yet been incorporated into the OpenFlow protocol specification. However, what is there is good enough to successfully build and operate a network.

When implementing SDN, Google chose not to control it from a single NOC. Instead, network control is replicated and distributed to boost the network's fault tolerance. Google views having replicated distributed control as a fundamental design requirement (Dix, 2012).

The importance of the centralized controller cannot be overstated. Considering the critical functionality that is provided by this network device, a backup version must always be available. In practical terms, this means that control plane functionality has to be implemented that will allow the various centralized controller boxes to elect who is currently the MASTER controller. Correctly implementing this logic is difficult and complex to do, requiring both effort and time.

Another important question that has to be answered is where software functionality should be located: on the switch or on the centralized controller. Google's plan is to move as much functionality to the centralized controller as possible, but some must still be left on the switch. What to leave and how to configure what has been left are inexact sciences.

In very large networks, there will be many different traffic flows. The occurrence of a major configuration change in the network means that most, if not all, of those traffic flows will now have to be recomputed. Initially, Google ran into performance problems when they tried to recalculate the traffic flows quickly enough not to have an impact on network performance. These problems were eventually solved.

The ability to introduce SDN so quickly was supported by the ability to write and introduce network software when it is being used in a standard environment on a powerful centralized server. This is in contrast to writing software for a router environment that is limited in terms of functionality and processing power.

7.9   Lessons learned

Having completed the conversion of the G-Scale backbone network to use SDN technology and retiring the legacy networking gear that had been used to provide the network, some conclusions have been reached about the use of SDN technology.

The first conclusion is that the OpenFlow protocol, although not complete, is ready for use in real-world networks (Hölzle, 2012). More functionality is needed; however, this should not hold network designers back from creating new networks based on the OpenFlow protocol.

The SDN architecture with its use of a centralized network controller, while supporting redundancy (Dix, 2012), is an idea that is also ready for production network use. Networks that are implemented using this new technology will permit the rapid deployment of rich feature sets. Network operators will also benefit from SDN's simplified network management operations.

The G-Scale network is a relatively simple network. If SDN technology is applied to a more complex real-world network, then the benefits to the network operator would be even greater.

In the short term, more standardization is needed. Specifically, larger firms are looking for ways to reduce the amount of time it takes to program a large network and more support for hardware configuration (Neagle, 2012).

Google is fully committed to the use of SDN within its enterprise networks. The mission critical interdata center G-Scale backbone network successfully runs on the OpenFlow protocol. This network, by traffic volume, is the largest production network. The implementation of SDN technology has resulted in improved manageability, and over time, it is expected also to provide improved cost savings.

Google measures the value of implementing SDN technology in several different ways. One is by improvements in the utilization of the network bandwidth. This is a potentially significant cost saving and will help to amortize the cost of the development of the new OpenFlow switches and the centralized controller software. This increase in utilization is being provided with potentially better guarantees of service even under network failure conditions because of the rapid recovery of the SDN environment. The expectation is that the G-Scale backbone network will be operated at the same level of service as the legacy network with much less effort (and cost) (Hölzle, 2012).

These savings should come about because of the reduction in configuration and monitoring that SDN will require. The ability to have the network automatically react to events by itself with little or no human interaction required should further reduce the costs of operating SDN. Savings can be measured in terms of unit cost per megabits per month of bandwidth delivered for a given SLA (Hölzle, 2012).

chapter eight

OpenFlow

8.1 Introduction

The OpenFlow protocol structures communication between the control and data planes of supported network devices (https://www.opennetworking.org; Azodolmolky, 2013; Hu, 2014; Open Networking Foundation [ONF], 2013; Bansal et al., 2013). OpenFlow has been designed to provide an external application with access to the forwarding plane of a network switch (or router). Access to this part of the router can be gained over the network, which allows the controlling program not to have to be colocated with the network switch.

Traditional networking protocols have tended to be defined in isolation, with each solving a specific problem and without the benefit of any fundamental abstractions. The result of this isolation has been the creation of one of the primary limitations of today's networks: complexity.

As an example of this complexity, to move a device from one location on the network to another location on the network, networking professionals must touch multiple switches, routers, firewalls, web authentication portals, and so on and update access control lists (ACLs), virtual local-area networks (VLANs), quality of services (QoS), and other protocol-based mechanisms (ONF, 2012) using network management tools that operate at the device and link levels.

In addition, when these types of changes are being made, the network topology, vendor switch model, and software version all have to be taken into account. The end result of this network complexity is that once a network is built, it often stays as it is so that nothing becomes broken.

The OpenFlow protocol has been created to solve the problems that legacy networking protocols have created. In the software-defined networking (SDN) architecture, OpenFlow is the first standard communications interface defined between the control and forwarding layers. OpenFlow allows direct access to and manipulation of the forwarding plane of network devices such as switches and routers, both physical and virtual (hypervisor based). Currently, no other standard protocol does what OpenFlow does, and it has been determined that a protocol like OpenFlow is needed to move network control out of the networking switches to logically centralized control software.

When the OpenFlow protocol is implemented, it is implemented on both sides of the interface between the network infrastructure devices and the SDN control software. To identify network traffic, OpenFlow uses the concept of flows based on predefined match rules that can be statically or dynamically programmed by the SDN control software.

OpenFlow allows network professionals to define how traffic should flow through network devices based on parameters such as usage patterns, applications, and cloud resources. OpenFlow allows the network to be programmed on a per flow basis. This means that an OpenFlow-based SDN architecture can provide extremely granular control, enabling SDN to respond to real-time changes at the application, user, and session levels. In today's legacy networks, routing based on the Internet Protocol (IP) does not provide this level of control because all flows between two end points must follow the same path through the network, regardless of their different requirements.

The OpenFlow protocol has been created to enable software-defined networks and is now the only standardized SDN protocol that permits direct manipulation of the forwarding plane of network devices. OpenFlow was initially applied to Ethernet-based networks; however, OpenFlow switching can extend to a much broader set of use cases. OpenFlow-based SDNs can be deployed on existing networks, both physical and virtual. Network devices can simultaneously support OpenFlow-based forwarding as well as traditional forwarding. This means that it is easy for enterprises and carriers to progressively introduce OpenFlow-based SDN technologies, even in multivendor network environments.

8.2   Overview of the OpenFlow switch specification

The ONF is a user-led organization dedicated to promotion and adoption of SDN (https://www.opennetworking.org). The ONF manages the OpenFlow standard. Version 1.1 of the OpenFlow protocol was released on February 28, 2011. The next version of the standard (version 1.2) was published in February 2012. Version 1.4 of OpenFlow was current as this book was written and was released on October 14, 2013.

The ONF OpenFlow protocol identifies the requirements of an OpenFlow switch (shown in Figure 8.1). An OpenFlow switch consists of three main components: an OpenFlow channel, a group table, and one or more flow tables. The flow tables and the group table are responsible for performing packet lookups and packet forwarding. The OpenFlow channel is used to communicate with an external controller. The external controller uses the OpenFlow protocol to manage one or more OpenFlow switches (Azodolmolky, 2013; Hu, 2014).

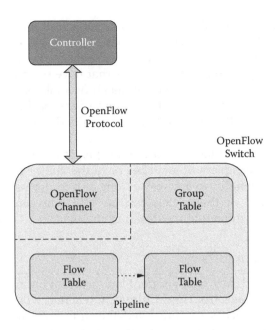

Figure 8.1 OpenFlow switch main components. (Adapted from Open Networking Foundation, OpenFlow Switch Specification, Version 1.4.0.)

In an OpenFlow switch, the abstraction at a high level is that Ethernet frames ("packets") arrive, and their headers are compared to data that has been stored in a flow table within the switch. Every OpenFlow switch must contain one flow table and can contain multiple flow tables. If a match between the contents of the packet's header fields and the flow table entry is made, then a set of instructions is then executed. Each one of the switch's flow table entries contains three pieces of information: the data that is used to match the fields in the received packets ("match fields"), counters that keep track of the number of matches that have been made, and the instructions that are to be executed if a match is successfully made.

The matching of the fields in a packet starts with the first table (numbered "0") and continues until a match is made or a table-miss event is declared to have happened. Exactly what happens when a table-miss event occurs is dependent on how the switch is configured, but options include dropping the packet, forwarding it to the controller for further processing over the OpenFlow channel, or continuing on to the next flow table to continue the search for a match to the packet.

As the network changes, the OpenFlow switch's flow tables have to be updated. Updating these tables is the responsibility of the external controller to which the switch is connected.

If a packet reaches the end of a flow table and it still has not been matched, then if the table-miss instructions modify the packet-processing pipeline, the packet may be allowed to be sent to the next flow table in the pipeline. Any time the instructions that have been retrieved from a flow table entry because of a packet match or a table-miss event do not specify a next flow table, then processing of the packet will come to a halt. The packet will generally then be modified and forwarded to the next switch.

The OpenFlow switch specification is just that—a specification. This means that the designers of switches have the ability to implement the OpenFlow functionality in any way as long as it conforms to the standard. The functionality of the switch can be split between hardware and software in any fashion that the designer chooses.

8.2.1   OpenFlow ports

Ports are a critical part of the OpenFlow protocol because they specify where a packet comes from and ultimately where it will be going. When two OpenFlow switches are connected, they are connected via ports.

An OpenFlow switch contains three different types of ports: physical, logical, and reserved. Each of these port types behaves differently. Any one of these port types can be used as both an input and an output port for a packet.

Physical ports are exactly what they sound like—ports that correspond to a physical interface port on the OpenFlow switch hardware. Logical ports are not related to a physical port on the switch. However, logical ports can be made to map to a physical port on the switch. When packets are being processed by the OpenFlow switch, both physical and logical ports are treated exactly the same way.

Reserved ports are special ports that are used to cause a specific action to occur. This action is triggered by sending a packet to a reserved port. An OpenFlow switch is required to support five types of reserved ports. There are three additional types of optional reserved ports that can be supported by the switch.

8.2.2   OpenFlow packet-processing pipeline

Figure 8.2 shows how the set of flow tables in an OpenFlow switch work together to process a received packet. Every switch must have at least one flow table. This table is assigned the number "0." Additional flow tables can exist, and each of them has an identifying number. A received packet will first have the contents of its header bits matched to the flow entries in table 0. If a match is made, then the instructions that are part of that flow entry will be executed. One of the instructions that is executed may direct

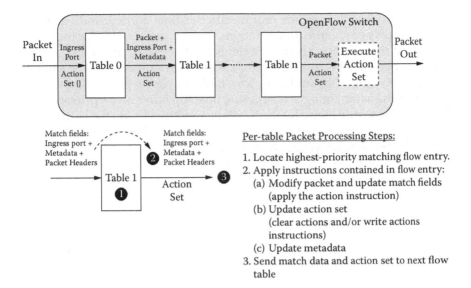

Per-table Packet Processing Steps:

1. Locate highest-priority matching flow entry.
2. Apply instructions contained in flow entry:
   (a) Modify packet and update match fields
       (apply the action instruction)
   (b) Update action set
       (clear actions and/or write actions
       instructions)
   (c) Update metadata
3. Send match data and action set to next flow
   table

Figure 8.2 OpenFlow packet flow through the packet-processing pipeline. (Adapted from Open Networking Foundation, OpenFlow Switch Specification, Version 1.4.0.)

the OpenFlow switch to send this packet to another table (that table must have a larger table number than the table that is currently processing the packet) to continue to attempt to make more matches between flow entries and the packet's header bits.

The packet-processing pipeline will stop when the flow entry that is matched to the packet does not have instructions that request that the packet be sent to another flow table for processing. Once this happens, the rest of the instruction set for this flow table entry will be processed against the packet, and then the packet will be forwarded by the switch.

Packets will not always match to the current contents of a flow table. When this occurs, it is called a table-miss, and it is handled as specified in the implementation of the OpenFlow protocol on a switch. Many different actions can be taken, including dropping the packet, passing the packet on to another table, or sending it to the controller via the control channel using *packet-in* messages (ONF, 2013).

8.2.2.1 Flow tables

At the center of an OpenFlow switch's packet-processing pipeline are its flow tables. These tables are used to determine what, if any, action should be taken based on receiving a given packet. The flow tables are an important part of the OpenFlow switch's packet-processing pipeline.

Each flow table consists of a number of flow entries, and each flow entry consists of six information items. Two of these items are the match fields and the priority. The match fields are the values that are compared against specific fields in a received packet to determine if there is a match.

It is possible that multiple flow table entries may match the same packet at the same time. If this occurs, then the flow entry's priority value is used to determine which match will be used to provide the instructions that will be executed against the packet. Figure 8.2 shows the steps that an OpenFlow switch goes though once a packet has been received.

If it exists, when the table-miss flow entry is matched to a packet, all fields are omitted from the comparison, and the table-miss flow entry has a priority of 0 (lowest priority).

8.2.2.2 Matching

Figure 8.3 shows the steps that an OpenFlow switch goes though once a packet has been received. Every time a flow table entry matches a packet, the counter associated with that flow table entry is updated. These counters are permitted to roll over. After the counters have been updated, then the instructions that are associated with that flow table entry will be executed.

In the unlikely event that two or more flow table entries match the packet and have the same priority level, then the OpenFlow switch will not be able to determine which flow entry's instruction set should be executed.

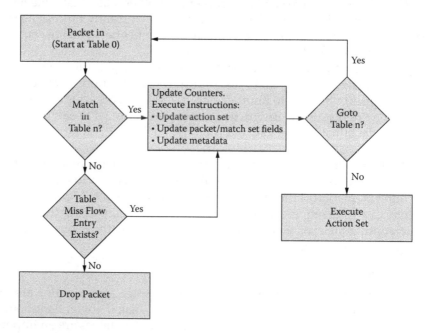

Figure 8.3 OpenFlow packet-matching process.

In this case, nobody will win. The OpenFlow switch will assume that the selected flow entry is left undefined, and the packet will not be changed.

8.2.2.3   Table-miss

A match between a received packet and a flow table entry will only occur if the flow table entry contains values that are the same as found in the header of the packet. If none of the entries in the flow table contains the values that are in the received packet, then a table-miss event has occurred.

The OpenFlow protocol states that every flow table must have a table-miss flow entry. However, the table-miss flow entry does not exist in the flow table by default. This means that the external controller is responsible for creating this table entry.

The table-miss entry will match every packet, and this match will have a priority of 0. When a packet is unmatched by any other flow table flow entry with a higher priority, the table miss entry will be the only match. When the match with the table-miss flow entry is the only match that is made, the table-miss flow entry instructions will be applied to the packet.

When a packet does not match any of the other entries in a flow table, the table-miss flow entry will determine how to process the packet. The actions that may be taken include dropping the packet, sending the packet to the controller, or sending the packet to another table. There are two actions that the table-miss flow entry must support. It must support at least sending packets to the controller using the reserved port CONTROLLER. In addition, it must support the dropping of packets using the OpenFlow Clear-Actions instruction (ONF, 2013).

8.2.2.4   Flow removal

SDN is dynamic. Links will go up and down all the time. This means that an OpenFlow switch's flow table needs to be constantly updated by the external controller. One key action that will be taken by the controller will be the removal of flows. Removals can occur at the request of the external controller, the switch flow expiration mechanism can cause a flow to be removed, or an optional switch eviction mechanism can be used to remove a flow.

The OpenFlow switch's flow table entry expiration mechanism works in one of two different ways depending on the configuration of the switch. A nonzero hard time-out counter seconds value can be associated with a flow that, when enough time has elapsed for it to count down to zero, will cause the flow to be removed. Alternatively, a second idle time-out value can be associated with a flow. If this counter has a nonzero seconds value, then if the flow does not match a packet that has been received by the switch before the idle counter reaches zero, the flow will be removed.

Flows can also be removed by the external controller. The controller accomplishes this removal by sending a message to the switch requesting that the flow be deleted. If the OpenFlow switch wants to reclaim switch resources, then it can evict one or more flow entries. Note that the ability to evict flows is an optional OpenFlow switch feature and must be enabled before it can be used.

The external controller can configure the switch to make it notify the controller whenever a flow is removed. This setting can be made on a flow-by-flow basis. Every flow removal message contains four pieces of information:

- A complete description of the flow entry
- The reason for the flow entry's removal (expiration, deletion, or eviction)
- The flow entry's duration at the time of its removal
- The flow entry's flow statistics at the time of its removal

8.2.2.5 Meter table

The OpenFlow protocol provided the ability to implement a simple QoS operated by using what are called "meters." Meters can be defined on a per flow basis. The OpenFlow protocol specification does not contain any required meter band types. There are only optional meter band types. To implement metering, the switch must allow the meter to measure the rate of the packets that are assigned to it. The goal of this metering is to allow the switch to control the rate of those packets that are being metered.

To implement the metered QoS features, the OpenFlow switch must first provide counters so that the rate of packets can be counted and measured. Eight types of counters are maintained. Counters are maintained for the following OpenFlow components:

1. Each flow table
2. Each flow entry
3. Each port
4. Each queue
5. Each group
6. Each group bucket
7. Each meter
8. Each meter band

The counters used by an OpenFlow switch can be implemented in software and then maintained by polling of hardware counters that have a more limited range that they can count. The counters listed in Table 8.1 are required counters that every OpenFlow switch must have.

An OpenFlow switch is not required to support all possible counters. The optional counters that an OpenFlow switch can support are shown in Table 8.2. For each counter, the duration refers to the amount of time that a flow entry, a port, a group, a queue, or a meter has been installed in the OpenFlow switch. Each counter must be tracked with per second precision.

Table 8.1 List of Required OpenFlow Switch Counters

Type	Counter	Bits
Per flow table	Reference count (active entries)	32
Per flow entry	Duration (seconds)	32
Per port	Received packets	64
Per port	Transmitted packets	64
Per port	Duration (seconds)	32
Per queue	Transmitted packets	64
Per queue	Duration (seconds)	32
Per group	Duration (seconds)	32
Per meter	Duration (seconds)	32

The Receive Errors counter is the total of all receive and collision errors associated with the counters in Table 8.1 and Table 8.2 as well as any others that are not identified by these tables. Each counter is unsigned and will wrap around with no indication that this has happened if its maximum value is exceeded.

If an OpenFlow switch does not provide a specific numeric counter, then its value must be set to the counter's maximum field value, which is the unsigned equivalent of −1.

8.2.2.6   Instructions

For a packet in SDN to get from its source to its destination, it will need to have its header modified by the OpenFlow switches through which it passes. The way that an OpenFlow switch determines how to modify a packet is based on the instructions that it executes against the packet. The collection of instructions that are executed will be based on matching the packet's header field to instructions that are contained in the flow table entry.

If for any reason an OpenFlow switch is unable to execute the instructions associated with a flow entry, then the switch must reject the flow entry. If this event happens, then the switch must return an unsupported flow error message. It is possible that OpenFlow switch flow tables may not support every instruction, every action, or every match.

8.2.2.7   Action set

In addition to the instructions that are contained in the flow table entries, each packet in an OpenFlow switch has an action set associated with it. Initially, this action set is empty.

Table 8.2 List of Optional OpenFlow Counters

Type	Counter	Bits
Per flow table	Packet lookups	64
Per flow table	Packet matches	64
Per flow entry	Received packets	64
Per flow entry	Received bytes	64
Per flow entry	Duration (nanoseconds)	32
Per port	Received bytes	64
Per port	Transmitted bytes	64
Per port	Receive drops	64
Per port	Transmit drops	64
Per port	Receive errors	64
Per port	Transmit errors	64
Per port	Receive frame alignment errors	64
Per port	Receive overrun errors	64
Per port	Receive cyclic redundancy check (CRC) errors	64
Per port	Collisions	64
Per port	Duration (nanoseconds)	32
Per queue	Transmit bytes	64
Per queue	Transmit overrun errors	64
Per queue	Duration (nanoseconds)	32
Per group	Reference count (flow entries)	32
Per group	Packet count	64
Per group	Byte count	64
Per group	Duration (nanoseconds)	32
Per group bucket	Packet count	64
Per group bucket	Byte count	64
Per meter	Flow count	32
Per meter	Input packet count	64
Per meter	Input byte count	64
Per meter	Duration (nanoseconds)	32
Per meter band	In-band packet count	64
Per meter band	In-band byte count	64

Depending on a match that is made with a packet, the corresponding flow entry can modify the packet's action set. The packet's action set travels with it as it moves between flow tables. The actions in the packet's action set will be executed when the instruction set of a flow entry does not contain an instruction that points to the next flow table and the pipeline processing of the packet has stopped.

8.2.2.8   Action list

The list of actions contained in a packet's action set is called an action list. The order in which the actions in the action list are executed is determined by their location within the action list. When they are executed, the results are immediately applied to the packet.

The way that an action list is executed is to start by executing the first action on the list on the packet and then executing each action on the list in sequence on the packet. The effect of executing these actions on the packet is cumulative. A copy of the packet will be forwarded to the designated port in its current state if the action list contains an output action. A copy of the packet in its current state will be processed by the appropriate group buckets if the action list contains group actions.

8.3   OpenFlow channel

An OpenFlow switch is connected to an external controller via an OpenFlow channel. This is the interface that the controller uses to configure and manage the switch, receive events from the switch, and send packets out of the switch.

The OpenFlow protocol supports the following three types of messages for exchanging information between the controller and the OpenFlow switch:

1. Controller to switch
2. Asynchronous
3. Symmetric

Controller-to-switch messages are used to directly manage or inspect the state of the switch and are initiated by the controller. Asynchronous messages are used to update the controller with network events and changes to the switch state and are initiated by the switch. Symmetric messages are sent without solicitation and are initiated by either the switch or the controller.

There are seven controller-to-switch messages that are initiated by the controller and may or may not require a response from the switch. The controller does not have to request that asynchronous messages be sent from an OpenFlow switch. Asynchronous messages are sent by the switch to controllers to denote a packet arrival or switch state change. There are three main types of asynchronous messages. Symmetric messages can be sent without solicitation in either direction. Four symmetric messages have been defined as a part of the OpenFlow protocol.

8.3.1   Message handling

The OpenFlow protocol does provide a reliable means for both message delivery and processing. However, the OpenFlow protocol does not automatically provide acknowledgments or ensure ordered message processing.

When a switch receives a message from the controller, it must process it, and depending on the type of message, the switch may generate a reply. An error message will be sent to the controller if the switch cannot completely process a message that it has received from a controller. It is important that the controller's view of the switch be kept consistent with the state of the OpenFlow switch.

8.3.2   OpenFlow channel connections

An OpenFlow switch and an OpenFlow controller exchange OpenFlow messages using an OpenFlow channel. In general, a single OpenFlow controller will communicate with multiple OpenFlow switches using multiple OpenFlow channels. A single OpenFlow switch will typically have either a single OpenFlow channel connection to a single OpenFlow controller or multiple OpenFlow connections to multiple OpenFlow controllers for backup and reliability.

An OpenFlow controller is generally located remotely and uses one or more networks to connect to a given OpenFlow switch. The only requirement of the controller/switch network is that it supports the Transmission Control Protocol/Internet Protocol (TCP/IP). The network that is used to support controller-to-switch communications can be a dedicated network, a shared network, or an in-band network (the network that is being managed by the OpenFlow switch).

The OpenFlow channel between the OpenFlow switch and the OpenFlow controller is generally a single network connection that uses the Transport-Layer Security (TLS) or plain TCP. It is possible to create an OpenFlow connection that is composed of multiple network connections to exploit parallelism.

The OpenFlow switch is responsible for establishing a connection with the OpenFlow controller. In some cases, the OpenFlow switch may permit the OpenFlow controller to establish a connection with it. However, in this case, the switch usually should restrict itself to using only secured connections (TLS) to prevent unauthorized access to the switch.

The port that is used to establish communication between the OpenFlow switch and the external controller must be a user-configurable, but otherwise fixed, IP address. This connection can be established using either a user-specified transport port or the default transport port.

Assuming that the switch has been preconfigured with the IP address of the controller to connect to, the switch will then initiate a standard TLS or TCP connection to the controller. Traffic both to and from the OpenFlow channel does not travel through the packet-processing pipeline. What this means is that the OpenFlow switch will need to identify incoming traffic as local before checking it against the flow tables.

A switch can detect that it has lost contact with all of the external controllers that it had been connected to when the switch detects echo request time-outs, TLS session time-outs, or other disconnections. When this happens, the switch must immediately enter either "fail secure mode" or "fail stand-alone mode," depending on the switch implementation and configuration. The difference between the two modes is that in fail secure mode, the only change to switch behavior is that packets and messages destined to the controllers are dropped. In fail secure mode, flow entries should continue to expire according to their time-outs.

One way that the switch and the controller may communicate is by using a TLS connection. The switch starts the TLS connection to the external controller when the switch is started. This connection is located by default on TCP port 6653. To authenticate both the switch and the controller, certificates signed by a site-specific private key are exchanged. It is required that each switch must be user configurable, with one certificate for authenticating the controller (the "controller certificate") and the other for authenticating to the switch (the "switch certificate").

Communication between the OpenFlow switch and the controller using a plain TCP connection is permitted. The switch will initiate the TCP connection to the controller on its startup. This TCP connection is located by default on TCP port 6653. The OpenFlow protocol designers recommend that when a plain TCP connection is used, alternative security measures to prevent eavesdropping, controller impersonation, or other attacks on the OpenFlow channel are also used.

8.4   Controller modes

The OpenFlow protocol supports three different modes for controllers that are connected to a switch. The roles are as follows:

1. EQUAL
2. SLAVE
3. MASTER

Figure 8.4 shows four scenarios for the connection of one or more controllers to an OpenFlow switch. In the first scenario, a single controller connects to the OpenFlow switch. The controller in this scenario is in the EQUAL role. When a controller is in this role, it will have full access to the OpenFlow switch. While in this role, the default action is that the controller receives all the switch asynchronous messages. This role permits the controller to send controller-to-switch commands to modify the state of the switch.

In the second scenario, two controllers that are both in the EQUAL role are connected to the same switch. OpenFlow switches can establish communication with a single controller, or they may establish communication with multiple controllers at the same time. If an OpenFlow switch has connections with multiple switches at the same time, then they will be considered to be more reliable and can continue to operate in OpenFlow mode if one controller or controller connection fails. When a handover between controllers is required, it will be handled by the controllers themselves, and there will be no required switch involvement. This provides two advantages: First, that it permits fast recovery from failures, and second, that it permits controller load balancing.

In scenario 3, a single OpenFlow switch is shown connected to three controllers: two in EQUAL mode and one in SLAVE mode. A controller has the ability to request that its role be changed. One way that a controller

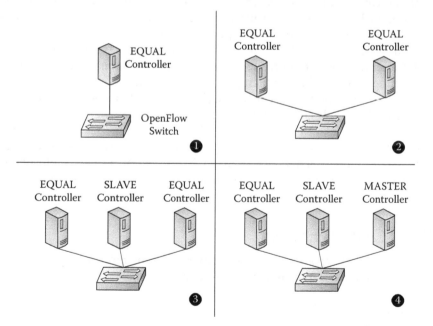

Figure 8.4 OpenFlow controller modes.

can cause this to occur is by requesting that its role be changed to SLAVE. This role only permits the controller to have read-only access to the OpenFlow switch. The default for this role is for the controller not to receive switch asynchronous messages, apart from port status messages. The controller is not permitted to execute all controller-to-switch commands that send packets or modify the state of the switch.

Scenario 4 shows an OpenFlow switch with one EQUAL, one SLAVE, and one MASTER controller. When a controller is in the MASTER role, the controller will have full access to the OpenFlow switch. The difference between this role and the EQUAL role is that, in this role, the switch will ensure that this controller is the only controller that is currently in this role. The switch is prohibited from changing the state of a controller on its own.

8.4.1   Auxiliary connection use for performance and reliability

When a controller is managing an OpenFlow switch, the performance of the channel between the controller and the switch may become a bottleneck. To alleviate this problem, the OpenFlow protocol permits multiple auxiliary connections to be established between the controller and the switch in addition to a main connection.

The OpenFlow channel traditionally consists of a "main connection." To boost the performance of the controller to switch communication along with the reliability of the connection, additional auxiliary connections between the controller and the switch can be created by the switch (Figure 8.5).

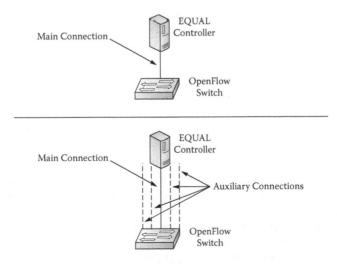

Figure 8.5 Auxiliary connections to an OpenFlow switch.

Each auxiliary connection must use the same controller IP address as the main connection. However, different transport protocols can be used on each auxiliary connection. These protocols can include TLS, TCP, DTLS, User Datagram Protocol (UDP), and so on.

Prior to establishing any auxiliary connections, the switch must first establish its main connection. If something happens to the main connection and it becomes unavailable to the switch, then all of the auxiliary connections that have been set up need to be taken down immediately.

8.4.2 Flow table synchronization

An OpenFlow switch may contain multiple flow tables as a part of its packet-processing pipeline. The switch may decide that a table in this pipeline should be synchronized with another table in the pipeline. What this means is that any additions or deletions to one table will be automatically replicated by the switch to the synchronized table.

The purpose of this type of synchronization is to permit the packet-processing pipeline to potently perform multiple matches on the same packet at different points in the pipeline. Switch implementation and configuration will determine exactly how tables will be synchronized because the OpenFlow protocol does not specify the details of how this is to be done.

8.4.3 Bundle messages

A controller will often have a need to make a large number of modifications to the configuration of an OpenFlow switch. There are two ways to go about making these changes. The first is to send each change message to the OpenFlow switch individually and have the switch make the change. Alternatively, a sequence of OpenFlow modification requests known as a "bundle" can be sent from the controller to the switch as part of a single message.

Bundles are handled uniquely by a switch. If all of the modifications contained in a bundle can successfully be made to a switch, then the modifications will be made, and the results will be retained by the switch. However, if for some reason an error is encountered while performing one or more of the modifications, the modifications will not be made and the results will not be retained.

The use of bundles allows a controller to synchronize its changes across multiple OpenFlow switches. Bundles can be sent to multiple switches, where they will be prevalidated. Once this has been accomplished for all switches, the changes can then be applied simultaneously.

8.5   OpenFlow configuration-and-management protocol

A separate configuration-and-management protocol has been created to manage the communication path that the OpenFlow controller and OpenFlow switch use to communicate. This protocol is called OF-CONFIG. The purpose of OF-CONFIG is to provide functionality that is not found in the OpenFlow protocol. The creation of the OpenFlow switch is assumed to have been done independently of the OpenFlow and OF-CONFIG protocols.

8.5.1   Remote configuration

The OF-CONFIG protocol can be used to remotely configure the OpenFlow switch's configuration and operation of the data path that connects the switch to one or more controllers. The OF-CONFIG identifies the various types (physical and virtual) of OpenFlow switches as OpenFlow logical switches. The components of the OF-CONFIG protocol are shown in Figure 8.6.

The OF-CONFIG protocol provides the ability to perform six functions:

1. The assignment of one or more OpenFlow controllers to interact with a given OpenFlow switch
2. The configuration of the queues and ports that an OpenFlow switch will use to communicate with one or more controllers
3. The ability to remotely change some aspects of an OpenFlow switch's ports (e.g., place them in an up/down state)
4. Configuration of security certificates for secure communication between the OpenFlow logical switches and OpenFlow controllers

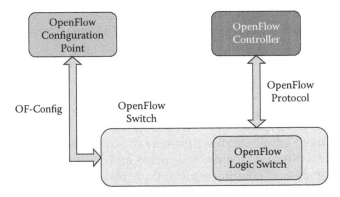

Figure 8.6 Relationship between OF-CONFIG protocol and the OpenFlow protocol.

5. Discovery of capabilities supported by an OpenFlow logical switch
6. Configuration of a small set of tunnel types, such as IP-in-GRE, NV-GRE, and Virtual Extensible Local-Area Network (VXLAN) (Bansal et al., 2013)

OP-CONFIG messages to the OpenFlow switch come from an OpenFlow Configuration Point. The OF-CONFIG specification does not detail exactly what an OpenFlow Configuration Point is.

8.5.2   Connection establishment between switch and controller

The OpenFlow protocol specifies that the OpenFlow switch set up its connection with the controller. To perform this action, the OF-CONFIG's OpenFlow configuration point must have configured the OpenFlow switch with the following three informational components:

1. The controller's IP address
2. The controller's port number
3. The transport protocol to be used by the connection (TLS or TCP)

Network devices that support OF-CONFIG need to implement the Network Configuration (NETCONF) protocol to be used as the transport protocol for OF-CONFIG.

The OF-CONFIG protocol provides the following functionality:

• The OF-CONFIG protocol specifies how to configure multiple instances of the parameter set for specifying the connection setup to multiple controllers.
• When a switch loses connectivity with a controller, it has to enter into the fail secure mode or the fail stand-alone mode. The OF-CONFIG protocol is responsible for configuring the parameters that the switch uses to determine which mode to enter when connection is lost.
• An OpenFlow switch has a number of queues that are used to process packets. Each queue has three parameters associated with it: min-rate, max-rate, and experimenter. The OF-CONFIG protocol is responsible for providing the information that is needed to configure these parameters for each queue.
• An OpenFlow switch has a number of ports that are used to communicate with other OpenFlow switches and controllers. OF-CONFIG is responsible for configuring four port-related parameters: no-receive, no-forward, no-packetin, and admin-state. OF-CONFIG also supports providing configuration information for port features and for obtaining state information on these port features.

- Every OpenFlow switch has a set of unique capabilities. The OF-CONFIG protocol provides a means for the switch to inform a controller which capabilities have been implemented on it.
- Each connection from the switch to the controller is identified by the switch Datapath ID and an Auxiliary ID. The OF-CONFIG protocol provides a means to configure the Datapath ID.

8.5.3 OF-CONFIG transport protocol

For an OpenFlow switch to support the OF-CONFIG protocol, it must support NETCONF protocol, an Internet Engineering Task Force (IETF) network management protocol. The switch's support of the NETCONF protocol provides the OpenFlow configuration point with the mechanisms to install, manipulate, and delete the configuration of OpenFlow switches. Its operations are implemented using a simple Remote Procedure Call (RPC) layer. The NETCONF protocol uses data based on the Extensible Markup Language (XML) encoding for the configuration data as well as for its protocol messages. The protocol messages are then exchanged on top of a secure transport protocol.

8.6 The conformance test specification for OpenFlow Switch Specification 1.0.1

The ONF has created a set of 206 test cases that can be used to determine the conformance of an OpenFlow 1.0.1 enabled switch to the OpenFlow protocol specification (v1.0.0 and the subsequent Errata v1.0.1). Currently, test procedures have not been created to validate security, interoperability, or performance features of an OpenFlow switch.

Some of the test cases contained in this test specification are mutually exclusive, optional, or only relevant for some implementations. In some of the test cases, the methods of validating the outcome of a test are not fully described and may be left up to the tester or test tool developer.

Three different profiles have been created to which OpenFlow switches can conform. These profiles are as follows:

1. Full
2. L2
3. L3

If a switch is to be considered fully conformant with the OpenFlow Switch Specification 1.0.0 and the subsequent OpenFlow Switch Errata 1.0.1, then for a given profile, the switch must satisfy the requirements of all test cases that indicate "MANDATORY" for "All" or the appropriate profile (Full/L2/L3).

The test bed used to perform the test cases is shown in Figure 8.7. The test bed consists of the following components:

1. A single test controller that has a single control channel connection to the device under test (DUT).
2. The test controller should be equipped with the ability to perform a packet trace and decode OpenFlow 1.0 packets.
3. There is a traffic generator/analyzer that has a minimum of four ports compatible with the DUT for data plane connections.

There are 10 groups of tests that have been specified to be performed to determine if a device is compliant with the OpenFlow protocol specification (Bansal et al., 2013). The 10 groups are as follows:

1. Basic sanity checks
2. Basic openflow protocol messages
3. Spanning tree
4. Flow modification messages
5. Flow matching
6. Counters
7. Actions
8. Messages
9. Async messages
10. Error messages

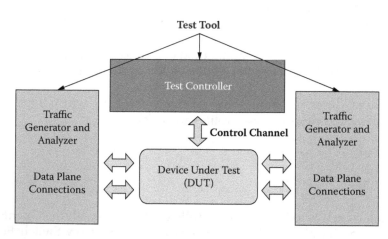

Figure 8.7 Test bed used to execute OpenFlow test cases.

8.7   The OpenFlow™ conformance testing program

The ONF has established an OpenFlow™ Conformance Testing Program to assist vendors in certifying that their hardware and software products meet the OpenFlow protocol specifications. The testing that is done as a part of this program is performed by accredited testing labs around the world.

When a vendor's product is able to pass the tests in this program, the product will receive a certificate of conformance. In addition, the vendor will then be permitted to use the ONF OpenFlow Conformance Testing Program logo as a part of its product-marketing efforts. Plus, certified vendor products will be listed on the ONF's website.

The ONF OpenFlow Conformance certification means that a vendor's product conforms to a specific version of the OpenFlow specification. Currently, the ONF offers three different types of certificates: Full Conformance, Layer 3 Conformance, and Layer 2 Conformance. The differences between these certificates are as follows (ONF, n.d.):

- **Full Conformance**: The tested device must be able to match all 12 "match fields" listed in the OpenFlow Switch Specification 1.0.0 (Errata 1.0.1).
- **Layer 3 Conformance**: The tested device must be able to match the following four fields in the OpenFlow Switch Specification 1.0.0 (Errata 1.0.1): Ingress Port, Ethernet Type, IP Source Address, and IP Destination address.
- **Layer 2 Conformance**: The tested device must be able to match the following five fields in the OpenFlow Switch Specification 1.0.0 (Errata 1.0.1): Ingress Port, Ethernet Source Address, Ethernet Destination Address, Ethernet Type, and VLAN id.

For a vendor to obtain ONF OpenFlow Conformance certification for a device, the vendor must first join the ONF. The next step is for the vendor to establish a contract with an ONF-approved testing lab (ONF, n.d.). That lab will then conduct a certified conformance test and, assuming that the tested product passes the tests, will certify it. Once this has been accomplished, the ONF will approve the awarding of an ONF OpenFlow Certificate of Conformance.

chapter nine

SDN evolution

9.1   Introduction

Software-defined networks (SDNs) are a new innovation in networking. The technology is emerging, implemented in a slowly growing community of real-world networks. There remains a great deal to learn about how to both build and manage SDN.

Researchers are using the available information about the OpenFlow protocol to create test networks and to run network simulations. Of particular interest is how SDN and legacy networks will interact. It is realized that, given the scale of networks today, there will never be a way to perform a "flash cut" to convert a legacy network into SDN overnight, so interoperability is going to be important.

The initial work on SDN has been to apply it to traditional L2/L3 enterprise networks. Researchers are only now starting to try to determine if SDN can be applied to other types of networks, such as optical transport and wireless. If it is possible to extend SDN to cover these additional networking technologies, then the next question will be to determine how that type of SDN environment can be managed.

SDN comes with a number of potential advantages over traditional networking technologies. The best way to maximize these benefits is unclear. Additional research is required to better understand how scalable this approach to networking is.

In a network, a middlebox is a computer networking device that transforms, inspects, filters, or otherwise manipulates traffic for purposes other than packet forwarding. The arrival of SDN means that a significant amount of current middlebox functionality may be able to be worked into the centralized controller. Questions remain about the best way to do this and how to accomplish the task without degrading the performance of the network.

The following sections identify open issues in SDN and discuss how researchers are investigating them. This is an active, ongoing research area that will evolve over the next several decades.

9.2   SDN and enterprise networks

This research tackles the challenging problem of how SDN technology can be added to existing enterprise networks. Much of the research into SDN has assumed that an end-to-end SDN could be built from the ground up. However, in the real world this is not possible. Existing legacy networks are going to have to be upgraded to use SDN technology. The mechanics needed to accomplish this are not known and have been the subject of ongoing research (Levin et al., 2013).

It is acknowledged that once a commitment has been made to transform an existing legacy enterprise network into SDN, the transformation will cause a hybrid network to be created. A portion of the network will consist of legacy networking equipment, and a portion will consist of new SDN equipment. Researchers want to understand if and how the two different networks can be made to work together.

In this networking scenario, three key questions that need to be answered have been identified:

1. Is there any benefit to the network owner to upgrade its existing legacy network to become a hybrid legacy/SDN environment?
2. In a hybrid legacy/SDN environment, how important is the actual placement of those switches that have been upgraded to SDN?
3. To obtain the greatest benefit from introducing SDN into the enterprise network, just how many of the existing legacy switches need to be upgraded to use SDN technology before benefits will start to be seen?

Figure 9.1 shows three possible ways that SDN equipment could be deployed into an existing SDN enterprise network (Levin et al., 2013). The dual-stack scenario in this figure shows SDN functionality (perhaps support for the OpenFlow protocol) being added to the legacy switches in the network. The result of this addition is that two separate side-by-side networks are created. When applications are connected over the network, a decision has to be made whether to send packets via SDN or via a legacy connection. A drawback to this approach is that it will be necessary to continue to deploy more and more switches with both SDN and legacy functionality to provide maximum value to the applications that chose to use both forms of connections.

The Access Edge scenario in Figure 9.1 shows an alternative way of introducing SDN technology into a preexisting SDN. In this scenario, the SDN functionality is only introduced at the edge of the network—in the access network. The core network will remain the domain of the legacy network switching gear. This approach would allow entire data centers that were part of the enterprise network to be virtualized using SDN technology. SDN will meet the legacy network when an SDN switch is

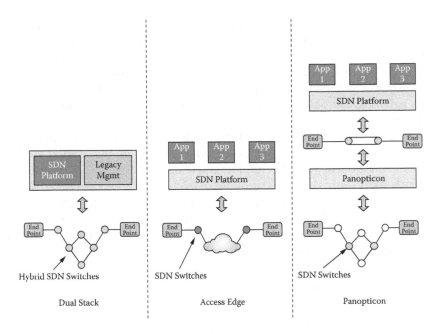

Figure 9.1 Three possible enterprise network SDN deployment scenarios.

interfaced to a legacy access switch. For the legacy access switch to be able to speak to the SDN switch, the legacy access software will have to be updated. In a typical enterprise network, there may be hundreds, if not thousands, of these access switches that will need to be updated. Clearly, the expense of implementing this solution will be great.

The final scenario shown in Figure 9.1 is called Panopticon. In this scenario, the SDN and legacy networks are integrated. The integrated network is what is shown to the SDN controller. The researchers report that from this scenario they have discovered that "the key benefits of the SDN paradigm can be realized for every source-destination path that includes at least one SDN switch" (Levin et al., 2013).

The conclusion that researchers have reached is that not every switch in the enterprise network has to be an SDN switch to obtain the value of deploying SDN technology. Instead, only a subset of the switches needs to be able to support SDN. As long as every path goes through at least one SDN switch, then a programmable network access control policy can be implemented. Additional functionality, such as load balancing, can be implemented for those network connections that pass through two or more SDN switches.

The Panopticon network architecture that researchers have created (Levin et al., 2013) ensures that every network connection will pass through at least one SDN node. The Panopticon architecture uses network waypoints to control network traffic and thereby implements the SDN abstraction.

A prototype of the Panopticon network architecture has been implemented, and many simulations of the network have been run. The results of these studies revealed that only an astonishing 0.6% of the switches in a typical enterprise network have to be upgraded before roughly 80% of the network can be operated as a single SDN. Note that, in this network, virtual local-area network (VLAN) and flow table updates can successfully be met.

Because the Panopticon network architecture allows the entire network to be viewed by the SDN controller as a single big virtual switch, end-to-end policies can be implemented on it. These types of policies include such things as application load balancing and access control.

What the Panopticon implementation has revealed is that a partial deployment of SDN may be all that some enterprise networks need to realize the SDN functionality that they want while living within the real-world constraints of limited network budgets and other resources.

9.3   SDN and transport networks

Researchers have studied the prices of the three major groups of equipment that are used in networks: processing and storage, routing and switching, and transport. They have discovered that processing and storage costs are dropping the fastest, and transport costs are dropping the slowest. Reasons for this are varied, but they include the fact that they are complex systems that deal with high-bandwidth communications, and there simply are fewer transport systems that are sold on average. What this means is that any changes to how a network is constructed that allows the existing transport network to be better utilized will provide significant benefits to the network operator.

The service providers who are in charge of transport networks are highly motivated to make their transport networks more sensitive to the changing bandwidth needs of the applications that are using them. In addition, it has been noted that during weekday business hours business users constitute the majority of transport network users. However, at night and on weekends (or holidays) residential users make up the majority of transport network users. The ability to reconfigure the transport network to better meet the needs of these two very different groups of users is an important goal for service providers.

The arrival of network virtualization services means that companies will soon have the ability to make use of external third-party providers of virtual private network (VPN) and cloud-based services. To make use of these services, transport facilities that connect the enterprise network to the external network will be required. The amount of bandwidth that will be required for these connections will vary and needs to be adjustable to meet current needs.

The ultimate goal is to be able to provide applications with the ability to specify both the bandwidth and the quality-of-service (QoS) parameters that they will require from the network. This request will need to be honored by the processing and storage components, the routing and switching network, and the transport network.

In trying to understand the need for introducing SDN technology into the transport network, researchers have studied several different scenarios in which transport networks are used. One such scenario involves the use of "cloud bursting," which occurs when an enterprise network elects to use external computing resources to augment its own computing power. This could occur during a busy holiday period or perhaps when year-end tax and other filings have to be created. Cloud bursting requires that the external network be initially configured to perform the tasks that have to be done. This generally requires that a great deal of data be transported from the enterprise network to the cloud-based network (McDysan, 2013).

Researchers believe that when smaller connections are being used to transport data (<1 Gbps), packet-switched networks will be able to efficiently handle the task. However, when the data transportation needs grow to become very large as during a cloud-bursting event (>10 Gbps), a better method is required. This method could be to bypass the packet network and directly use the transport network to connect the enterprise data center to the external data center. An OpenFlow-based SDN can be used to implement this type of solution. The goal will be to improve the efficiency of the transport network and to avoid any occurrence of blocking in the network.

This may require rearrangement of the existing transport network to permit it to support and work with the SDN (McDysan, 2013). Ultimately, the SDN controller will have to work with both the SDN components of the network and the legacy packet and transport network elements.

One advantage of incorporating the transport network into the overall network SDN processing is that improvements in network efficiency can be achieved. The routing and switching components of the network will be rapidly reconfigured to support the changing needs of applications. It is anticipated that the transport portion of the network would be reconfigured at a much slower rate to avoid packet reordering and potentially packet loss (McDysan, 2013).

One of the differences between the networks that use transport elements and other enterprise networks is that the transport networks often extend over multiple domains. Current interdomain routing and signaling standards do not support the ability to reconfigure the transport network to support the type of end-to-end optimization mentioned here (McDysan, 2013). This is an area of networking that still requires more work.

The arrival of SDN will change networking forever. This means that the configuration and use will also be changing. Use of the OpenFlow protocol will allow an abstraction of the physical network to be constructed and will ultimately result in greater simplicity when it comes to managing and configuring the network. To incorporate the transport networks that are already a part of today's enterprise networks into this new SDN future, change must be made.

Future research will investigate how to best provide bandwidth-on-demand features as multilayer networks are integrated with cloud networks. Prior work has focused on applying SDN technology to switching networks; the next steps include work to determine how to use OpenFlow to optimize how the transport network supports an application's bandwidth needs.

9.4 SDN and optical transport networks

In an ideal world, applications could be completely independent of the optical networks that they use to communicate. However, because of real-world communications issues such as latency, bit rate, and packet loss, applications have to be able to ensure that the network will be able to meet the application's needs.

Applications must be able to communicate with the network's control plane to establish the types of network routes and to identify the network resources that will meet the application's communication needs. Transport networks represent a special case when communicating with the control plane. Because of the distances that they cover, transport networks often have multiple segments, each of which may have its own technology and administration. To provide applications with the services that they need, interoperability between these various segments will be required.

In addition, in today's transport networks, many of the administrative functions are performed manually. The arrival of SDN is expected to automate many of these functions (Gringeri et al., 2013). The hope is that many of today's manual processes that can be time consuming and difficult to perform can now be automated to optimize how network resources are being used.

The OpenFlow protocol does not currently provide the ability to configure wavelength or circuit-based equipment. Current transport networks are generally controlled by element management systems (EMSs) or network management systems (NMSs), which are designed to permit manual control and configuration of the transport network. These management systems can be enhanced by adding a programmable interface (application programming interface, API) that will permit them to be externally controlled by another application. It is through an interface like this that the SDN control plane may be able to control an optical transport network.

One of the biggest challenges that transport networks face is that their connections can have specific performance characteristics. These characteristics can include, but are not limited to, bandwidth, connectivity, QoS, and resiliency. These parameters can change over a period of time, such as workday requirements being different from evening or off-hours requirements. If there was a way to allocate transport network resources dynamically, then the network operator could realize improved network utilization.

There are many different ways to create an optical transport network architecture. Each configuration can potentially be managed as part of SDN. Within a typical transport network, the switching can be performed based on packets, time slots, or fiber. Wavelength switching is a relatively new form of transport network. Wavelength ("color") paths can be switched; however, the process is slow and can take several minutes to complete.

Within a transport network, the control plane can be either distributed or centralized. If a single network end point is to be connected to multiple network end points, it is possible that a connection can be set up; however, it may not be an optimal connection. Links may be over- or underloaded, paths may be too loaded to accept another link, or a connection cannot be made because the network topology is not known.

Many of these problems can be solved using a centralized controller. In a transport network, a centralized controller has the ability to compute an end-to-end path that satisfies an application's connection, circuit, or wavelength constraints (McDysan, 2013). As in packet SDN, a centralized controller in a transport network is able to create links that are more efficient. When a path is needed as part of a high-priority connection, existing connections on that path can be rerouted to other paths, and the final network configuration can still meet the performance needs of all connections involved. Diverse routes for connections that share the same set of end points can also easily be calculated.

A centralized controller may not always be the ideal solution for a transport network. It is believed that a centralized controller may be slow to react to a link outage. In this case, a distributed controller that was running on the same processing platform as the fiber switch would be able to react more quickly and minimize packet loss.

The scale of transport networks makes them appear different from most packet networks. This may require a different architecture for the control plane that will be used with transport networks. A single unified controller may be made up of a series of geographically distributed controllers. These controllers will be responsible for managing their portion of the transport network as well as for communicating with the other controllers. Taken together, the multiple controllers will create a single unified controller for the entire transport network.

Transport networks often cross service provider and geographic boundaries. This means that different parties will be responsible for controlling different parts of the same transport network. Interfaces between the different controllers will be required to provide applications with end-to-end connection management. Researchers believe that this can be handled by using simple APIs or existing mature protocols (McDysan, 2013).

Just one type of controller will not be enough for transport networks. Instead, both centralized and distributed controller functionality will be required. The centralized controller will be responsible for coordinating an application's requirements with the network; the distributed controller will be responsible for providing a fast reaction to network events, such as a link failure. It is believed that an embedded controller may be more scalable and can be built on existing mature network protocols.

Transport networks are one part of a multilayer network. Adding an SDN controller to such a network provides an opportunity for the optical transport network to be better utilized to connect network applications. With the appropriate programming, an SDN controller could configure the optical transport network by setting the modulation scheme, symbol rate, forward error correction (FEC) overhead, or other parameters (McDysan, 2013). Proper control of the transport network could require that the controller have knowledge of vendor-specific equipment characteristics and because of this a two-tier controller structure, centralized and distributed, may be the optimal design. Researchers believe that the two separate controllers may operate on different timescales.

9.5   Increasing WAN utilization with SDN

The number of data centers is increasing, and their importance is growing at the same time. Global service providers use a large number of data centers as a key part of their service offerings. These firms have constructed separate wide-area networks (WANs) to interconnect these data centers.

These WANs are expensive to build and to operate. The incredible bandwidth that is required comes with an equivalent large price tag. Unfortunately, studies have shown that the average utilization of the links in these expensive WAN networks is generally 40–60% at best (Chi-Yao et al., 2013). The reasons for this low utilization are many and varied, but the fact that applications are unable to communicate when they need the network and how much bandwidth they will need over time results in each connection reserving the maximum amount of bandwidth that it will need for the entire duration of its connection.

One of the key reasons for the poor utilization of WAN links is because no global view of the network exists. In the absence of this view, routers will make decisions that are optimal for their local network environment. The end result will be a suboptimal overall network configuration.

The software-driven WAN (SWAN) (Chi-Yao et al., 2013) was created to overcome poor link utilization. This is a system that has been designed to allow a WAN to better utilize its links using SDN technology along with a controller that maintains a global view of the network.

Any SDN that wants to provide high utilization has to perform rapid updates of each switch's flow tables as the needs of the network applications change. However, this introduces problems into the network. As the updates are made, transient network conditions can be created. The changes have to be made to multiple switches in the network. The time that exists between when one switch has been updated and when the other switches have been updated allows network congestion to occur. This congestion can then have a negative impact on any applications that are latency sensitive.

The SWAN system addresses the update/congestion problem by reserving a portion of every path in the network (roughly 10%) for use by it to provide switch update information ("scratch capacity") in a nonblocking fashion. Another challenge that the researchers encountered was that the number of forwarding rules that a given commodity SDN switch would need to support would far exceed its forwarding rule capacity. The researchers addressed this issue by creating a system that would dynamically change a switch's forwarding rules based on traffic demand.

Not all traffic in a data center environment is the same. Service operators identified three primary types of services that a data center WAN would need to support (Chi-Yao et al., 2013):

1. **Interactive**: An interactive service is a request by one data center to another data center to obtain information that is needed to provide an end user with a reply. A characteristic of this type of data center traffic is that it is sensitive to both loss and delay.
2. **Elastic**: An elastic service is one in which two data centers exchange information to ensure that accurate answers can be provided to queries that are sent to an application in a given data center. This data has to arrive at the other data center within seconds or minutes. The impact of not getting the data there fast enough depends on the application that is using the data.
3. **Background**: A background activity is part of a data center's maintenance and provisioning tasks. This can include copying large data sets between data centers for backup or to start an application at a different site. This traffic requires a large amount of bandwidth and does not have a fixed time by which it must complete; however, completing it as quickly as possible is desired to minimize costs.

The amount of data of each type that data centers exchange is not equally distributed. Interactive data is the smallest amount, followed by elastic, which is then followed by background in size.

In today's data center networks, the traffic is engineered using existing network protocols such as Multiprotocol Label Switching–Traffic Engineering (MPLS-TE). This protocol creates a set of tunnels between two points in the network, and then the traffic between those two points is spread over the various tunnels. The priority of the data is determined by assigning a priority to a tunnel and then mapping traffic with that priority to that tunnel. In addition, each packet contains a differentiated service packet code (DSPC). This can be used to map a packet into a priority queue.

The MPLS system has two problems associated with it in real-world networks: efficiency and sharing. Poor link efficiency is caused by applications in networks not being aware of the bandwidth needs of other applications. Applications request the maximum amount of bandwidth needed independent of any other application's needs at the time. These types of requests can lead to network over- and undersubscription. Also, because a router using the MPLS protocol can only "see" the local environment, the router will make locally greedy decisions when trying to create paths. These local solutions may turn out to be inefficient.

Not all services that operate in a data center environment are equal. Some services should take priority when requesting network resources. Applications tend to obtain throughput based on rate that they are sending packets (Chi-Yao et al., 2013). This is a nonideal outcome. Attempts to map services into queues with their shared priority often do not work out because there simply are not enough queues to support all of the services.

The SWAN system has been designed to perform two functions: carry more traffic and support network-wide sharing of networking resources. SWAN supports three types of traffic priority classes: interactive, elastic, and background. To permit congestion causing updates of the routers in the network to occur, SWAN uses a multistep update process. This allows the SWAN system to guarantee that nonbackground traffic will experience no congestion, and background traffic will only encounter congestion that can be bounded.

A significant limitation of SDN routers is that, in real-world networks, they are required to maintain a large number of forwarding rules (perhaps thousands). Most SDN switches do not provide support for that many routing rules. The researchers have devised a method as a part of SWAN to dynamically change the rules and ensure that only the rules that are currently being used are in the switch at any moment in time.

The SWAN architecture consists of three main building blocks: service brokers, network agents, and a controller. The controller is responsible for coordinating all activity in the network. In the network, those services that are *not* interactive will be assigned a service broker with the responsibility to aggregate bandwidth demands from host applications and then provide them with the bandwidth that they need.

The network agents exist between the controller and the network switches. The SWAN network architecture is shown in Figure 9.2.

Here are the details on the various roles that each component plays in the SWAN architecture:

Service hosts and service brokers: The service brokers have the responsibility of estimating the bandwidth needed by the various service hosts and then limiting requests to the maximum bandwidth that the controller has allocated to the service broker. Each of the service hosts makes a request for the amount of bandwidth that will be required for the next 10 seconds. The service broker updates the controller with the aggregated demand from the service hosts that it is connected to every 5 minutes.

Network agents: With help from the network switches, the network agents are responsible for tracking changes in the network's topology and its traffic. Changes in the network's topology are reported to the controller immediately; traffic measurements are reported to the controller every 5 minutes. The network agent has the job of making sure that the OpenFlow flow tables in each of the switches that it is responsible for are correctly updated.

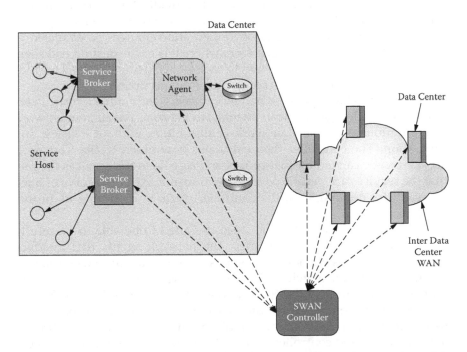

Figure 9.2 Architecture of SWAN. (Adapted from C. Hong et al., *Proceedings of the ACM SIGCOMM 2013 Conference*, 2013, Vol. 43, No. 4, pp. 15–26.)

Controller: This network component has the responsibility of performing three tasks every 5 minutes:

1. Calculate the amount of bandwidth that should be allocated to the various network components. This includes each of the service brokers and the interactive applications.
2. Notify services when bandwidth allocations have been decreased. Once this task has been performed, the controller will wait a fixed amount of time for the other party to implement the decreased bandwidth usage.
3. Change forwarding states of switches and then communicate to services with bandwidth that has been increased.

One of the most important features of the SWAN architecture is the ability to make many network updates without introducing congestion into the network. To make this happen, care is taken when performing two types of updates: traffic distribution across tunnels and when updating tunnels themselves.

When the controller is making updates to how it distributes network traffic across the existing tunnels, the controller can easily run into a congestion situation when links are heavily loaded. However, because the SWAN architecture requires that some bandwidth be reserved on each link ("scratch bandwidth"), this bandwidth can now be used as a part of the traffic distribution. A series of configuration changes is calculated to move the network configuration from where it is to the desired end state. These states are then implemented in each switch using the scratch bandwidth to prevent any network congestion from occurring.

Likewise, when updating network tunnels, the controller calculates a sequence of changes that can be made to get from the current network configuration to the new network configuration without causing any congestion. The controller then calculates how much traffic from each service can be carried by that tunnel and signals to the services the rate that the services are permitted to use to transmit. After the services have changed the transmission, the controller starts to distribute its tunnel changes.

The SWAN architecture has been designed to be able to gracefully handle network failure conditions. The network agents that interface directly with the switches are responsible for detecting switch failure conditions. When a switch or link failure is detected, the network agent will communicate it to the SWAN controller. On receiving notification of a failure, the controller will then compute new allocations for the network. All three major components of a SWAN network, service broker, network agent, and controller have backups. In the event that one of these components fails, the backup will take over. The backups are not "hot" backups—they do not maintain a copy of the current state of the network.

Instead, on taking over, they will query the other network components to build a map of the current state of the devices that are attached to them.

To study the level of optimization that the SWAN architecture could deliver, a test bed was constructed that consisted of five data centers in three different continents. Each one of the data centers had two WAN-facing switches and five servers per data center (Chi-Yao et al., 2013). The researchers compared the SWAN architecture to that of an MPLS-TE solution. The MPLS-TE solution was able to carry approximately 60% of the offered traffic. The SWAN architecture, however, was able to carry 98% of the traffic. This means that the SWAN architecture was able to carry 60% more traffic than the MPLS-TE solution. Ultimately, this was the goal that the SWAN researchers had designed the SWAN solution to accomplish.

9.6   How scalable are software-defined networks?

SDN permits the control plane to be separated from the data plane. This allows both of the sets of devices in these two planes to evolve separately. However, SDN comes with a unique set of concerns. With a single control plane for an entire network, important questions about the control plane's scalability and its performance have arisen. Test bed networks are a good way to test and evaluate protocol features; however, what will happen when SDN is applied to very large real-world networks?

9.6.1   SDN scalability issues

Networks grow in many different ways. Switches are added, available bandwidth increases, and more flows can be handled. A centralized controller is not believed to be able to grow at the same rate as the network. This may result in the controller becoming a bottleneck in the network. This concern was heightened by early test results that were based on the first SDN controller, NOX. It was reported that the NOX controller could only process 30,000 flow initiations per second (Yeganeh, Tootoonchian, and Ganjali, 2013), with each flow initiation supporting a flow installation of less than 10 milliseconds. The concern in the SDN community was that 30,000 was the limit for SDN size.

9.6.2   Controller designs for scalability

One SDN configuration consists of a single SDN controller providing service for the entire network. Clearly, as the network grows, the potential arises that the controller can be overwhelmed with network events and bandwidth requests. No matter how powerful the server the controller is running on is, there will be a point when the controller is not able to meet the needs of the network. Researchers point out that a data center environment may quickly reach this point with a high flow initiation request rate.

To improve the controller's ability to scale with the network, there are three changes to the controller that can be made. The first is to optimize the input/output (I/O) processing of the controller. This can include adding I/O batching to minimize the overhead of I/O and porting the I/O handling harness to the boost asynchronous I/O (ASIO) library (which simplifies multithreaded operation). In addition, the controller code is made aware of the environment in which it is running by using a fast multiprocessor-aware malloc implementation that scales well in a multicore machine (Tootoonchian et al., 2012).

Next, the load on the controller can be minimized by reducing the number of requests that the network forwards to the controller. One way to do this is to have smaller flows handled by custom hardware and only send the larger flows to the controller for processing. This reduces the load on the controller, although fine-grained flow-level visibility is surrendered to gain controller scalability.

Finally, the controller itself can be distributed over multiple systems. Within an SDN environment, several servers can fulfill the role of controller. This requires that there is a single unified view of the network. However, it should be noted that the more consistent the multiple controllers attempt to be, the greater the delay introduced into the system will be. Relaxing the requirement that each of the multiple controllers has to be consistent with the other controllers at all times will reduce the overhead that is required to make this happen.

An alternative approach to distributing the controller consists of a two-part controller instead of multiple controllers. One part of the controller is a locally scoped application deployed close to the switch. This application is then responsible for processing frequent requests that are exclusively local in scope. By doing this, a portion of the processing load that would be put on the central controller is removed. This two-part controller solution also has a centralized controller. The centralized controller handles any requests that are network-wide while helping to arbitrate between locally scoped applications.

Researchers believe that the scalability issues that are faced by SDN are not unique to SDN (Yeganeh, Tootoonchian, and Ganjali, 2013). Both the convergence and consistency issues that SDN faces are found in other protocols. The SDN differences are seen in the following two observations:

1. SDN allows the network designer to make a set of design trade-offs based on the constraints imposed on the SDN control program design.
2. Software development and standard distributed system design techniques can be used when designing a SDN controller. This allows standard software development practices to be used to verify and debug the controller design.

SDN frees the network designer from issues that a traditional network protocol designer has to repeatedly resolve. These issues include resiliency, distributing state information to parts of the network, and discovering the topology of the network. The availability of this information makes it easier to create applications that will eventually run on the controller platform.

9.6.3   Potential SDN scalability issues

In SDN, there are two types of flows: reactive and proactive. Reactive flows are created when a switch receives a packet for which the switch does not have a corresponding flow table entry for processing. When this happens, the switch notifies the controller, the controller provides the switch with an update to its flow table with which to handle the packet, and then the packet is processed. This approach allows for fine-grained, high-level, network-wide policy enforcement. In a proactive scenario, the controller updates the switch with the new flow table information before the packet is received, which allows the controller to avoid flow setup delay penalties.

In SDN, the flow setup process has four steps (Yeganeh, Tootoonchian, and Ganjali, 2013), which are shown in Figure 9.3:

1. An SDN switch receives a packet. On checking the packet's header information, the switch determines that it does not match any entry in the switch's flow table.
2. The switch sends a request to the controller to have a new flow table entry created for the received packet.

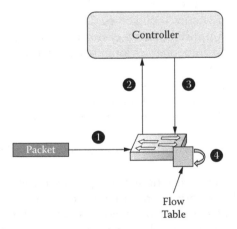

Figure 9.3 SDN flow setup process.

3. The controller provides the switch with the data that is to be placed into the flow table's new forwarding entry.
4. The switch uses the new information from the controller to update its flow table.

Analysis of the delays that are imposed on the flow setup process by each step identified that a controller running on a commodity server should be able to respond to a flow setup message from an SDN switch within a millisecond even when the controller is handling several hundred thousand such requests per second (Yeganeh, Tootoonchian, and Ganjali, 2013).

Controller designs are expected to keep pace with SDN growth to support reactive flow setup. The design of the control logic will be what ultimately determines the scalability of the controller. One other problem with SDN is that as the number of flows in the network increases, the flow table size in each of the switches may be exceeded. The ability to install aggregate flow table rules that match a large number of very small flows to reduce the load on the controller is being studied to address flow table size management.

Every network has the challenge of determining how to react to failures in the network and how much time is needed to recover from a failure, a process called *convergence*. SDN also has failure response concerns. Initial SDNs used a single controller, and this only served to make the resiliency issue an even bigger deal.

If a controller in an SDN environment fails, the system must determine the recovery method to use. Other controllers must discover that the controller has failed, and as long as there is a state-synchronized backup controller, there will not be any network state data lost. When a controller fails, the connection to the switches that the controller is managing will be terminated by the switches themselves. This means that the switches will need to be able to support a discovery mechanism that will allow them to discover the backup controller as long as the backup controller is within the switch's partition.

In SDN, network failure response differs from the response mechanism of a legacy network. The following five steps, shown in Figure 9.4, are executed on failure (Yeganeh, Tootoonchian, and Ganjali, 2013):

1. The SDN switch will detect that a network failure condition has occurred.
2. The SDN switch will send a message to the controller notifying the controller that a network failure has occurred.
3. Once the controller has been notified of the failure, it will use its knowledge of the network to compute new flows that do not use the failed component.

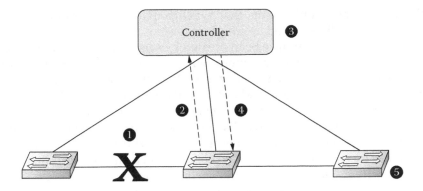

Figure 9.4 SDN convergence on a link failure.

4. The updated flow table information will be pushed out to the affected switches by the controller.
5. The SDN switches that are affected by the failure will receive the updated flow table information from the controller and will then use the information to update their flow tables.

One of the major differences between SDN and a legacy network is that, in a legacy network when a failure occurs, information about the failure is flooded out to the network. In SDN, the information is only sent to the controller. This means that the time needed to notify the network that a failure has occurred is the same in both legacy and SDN environments.

In SDN, the failure recovery process is similar to that of a legacy network. Both networks share the same scalability issues. Similar techniques can be used in both types of networks to minimize network downtime.

9.6.4 Network types

Different types of networks pose different challenges for SDN technology. Data centers represent a significant challenge for SDN. A single data center that supports virtualization can contain tens of thousands of switching elements. This represents a threat to any centralized controller because the number of control messages that could be generated could easily overwhelm a controller.

To prevent this from happening, one solution is to install rules on switches in the data center before the rules are needed. Doing this will eliminate the situation in which a switch receives a packet that it does not know what to do with and eliminates the need for the switch to contact the controller. A data center is an ideal network for supporting a distributed controller because of the availability of computing resources.

The ample bandwidth means that synchronizing the network views of multiple controllers should be simple to do.

The networks that have been created by service providers look much different from those of data centers. Service provider networks are not as equipment dense as a data center network is, and a service provider network will potentially be spread out over a long distance. Just the size of a service provider network can cause concerns about introducing latency in the updating of a centralized SDN controller. Generally, such a large network could be broken up into separate regions, with each region distinctly managed by its own controller. The collection of controllers can then synchronize their network views to create one common network view.

It is assumed that a service provider's network would contain a large number of flows. To permit the SDN switches to manage a large number of flows, flows are recommended to be aggregated by the switches to minimize the number of table entries that have to be supported (Chi-Yao et al., 2013).

9.6.5 Next steps

SDN provides network designers with both opportunities to perform network functions that have never been done before and challenges that will have to be overcome before the true benefits of using SDN technology can be realized. For a controller to manage a complete network, a great deal of low-level network information has to be provided to it via the OpenFlow interface. The applications to be built on top of the controller do not need to be exposed to all of the low-level information. Instead, the applications can operate using the abstracted view of the network that the controller will create and present to them. Hiding the details of the network implementation will allow the higher-level applications to scale with the network.

To correctly implement and manage SDN, network designers must have the ability to test different SDN configurations. The use of a single centralized controller greatly simplifies the process of testing the network under different configurations to determine how it will react. Any SDN testing tools that are developed will need to be able to operate on networks of varying scales. This is an area of research that will be active for a long time.

SDN will eventually have to support a wide variety of other networking protocols. The APIs that are used in SDN will have to be modified to support additional protocols (Yeganeh, Tootoonchian, and Ganjali, 2013).

The conclusion is that SDN does not pose any more of a scalability issue than legacy networks do. It is believed that the scalability issues that SDN does have can be resolved without taking away any of the novel functionality that SDN offers.

9.7 SDN management

The arrival of SDN provides support for network monitoring and control functions that have never before been available to network operators. The primary protocol that is used in SDN is the OpenFlow protocol, which allows a centralized controller to control network switches at a granular level. From a network control point of view, this is a powerful feature. However, programming the network becomes much harder because of the large number of operations that need to be executed and the high probability that some important step might be missed. A better way to program SDN is needed.

The OpenFlow protocol provides the network programmer with the ability to manipulate the state of the network devices. Creating programs that simultaneously modify how the network performs routing and access control is challenging. This task becomes even more difficult when one takes into account the additional challenges that must be addressed when the network switches are asynchronously updated with changed data. Researchers have labeled the writing of applications for today's SDN as "a tedious exercise in low-level distributed programming" (Foster et al., 2013).

The Frenetic research project has been created to design simple abstractions for programming SDN (Foster et al., 2010, 2011). The goal of Frenetic is to provide network programmers with the ability to easily control the three main stages of SDN management:

1. Monitoring of network traffic
2. Specifying and creating packet-forwarding policies
3. Updating network policies in a consistent way

Frenetic accomplishes this by offering a set of declarative abstractions. These can be used by the programmer to query the state of the network, define network forwarding policies, and update the network policies in a consistent manner (Foster et al., 2013). An important characteristic of the Frenetic approach is that it permits individual policies to be created in isolation. Once created, these policies can then be combined with other policies to create sophisticated network instructions.

Frenetic views programming the network as a three-part "control loop." The first part of this loop is where the network state is queried. Applications are provided with the ability to subscribe to updates on the state of the network. The methodology for creation of the updates is left to the runtime system. Applications can be provided with traffic statistics and topology changes that are based on the results of the runtime system collecting switch OpenFlow counters, switch statistics, and event messages.

The next step in the control loop is the reconfiguration of the network based on the state information that has been provided to the application. Ultimately, the network application will specify what packet-forwarding behavior a switch should exhibit; however, the way that it does it is via a high-level policy language that is provided by Frenetic.

The final step in the control loop is the reconfiguration of the network. Again, abstractions are used to permit the network application to be able to update individual switches without having to require that the network programmer specify the contents of each OpenFlow flow table entry on each switch. It is the responsibility of the runtime system to ensure network consistency and make sure that the network uses either the old network configuration or the new network configuration but never a combination of the two.

9.7.1   Network state queries using Frenetic

There are many different events that can change the state of an SDN environment. These include link failures, topology changes, changes in the network traffic load, or even the arrival of a specific packet at a specific switch. The SDN controller is able to monitor the state of the network by polling switches and retrieving the packet counter values associated with each flow table entry. Clearly, with even a single switch, there could be a great deal of data that must be monitored. The network programmer does not have to worry about this as it will be taken care of by the runtime system. The network programmer is free to focus on determining what to monitor without having to worry about how it will be done. Frenetic provides network programmers with the tools that they need to control what information they have. Frenetic provides four high-level operators (Foster et al., 2013) to accomplish this:

1. Classifying
2. Filtering
3. Transforming
4. Aggregating

Switches in SDN have a limited amount of space to store packet-forwarding rules. This means that it is not possible for the network programmer to configure how a switch should react to every type of packet that it might receive before packet processing starts.

Instead, the Frenetic system allows the network programmer to specify how a switch should react when it receives a packet with a given header. The runtime system will store this information. When a switch does receive a packet with the header, the switch will ask the controller what to do with the packet. The runtime system will instruct the controller to respond with the data provided by the network programmer. The switch will place that information into its limited forwarding rules table and process the packet using it.

Frenetic allows the network programmer to easily collect network statistics. Instead of forcing the programmer to regularly request that switches be polled to receive updated statistics, Frenetic allows the network programmer to specify an interval for statistics collection. The runtime system will then take care of the task of collecting and presenting the statistics to the network application.

9.7.2 Network policy creation using Frenetic

Network policies dictate how the network performs tasks such as monitoring, access control, and routing. One of the most significant challenges associated with network policies is when they are created by different individuals or applications and then applied to the same network. Conflicts can occur. Frenetic has developed a network policy language that has a number of features created to allow policies to be created and combined in a modular fashion (Foster et al., 2011).

When an SDN switch is powered up, each time it receives a packet the switch will not know where to forward the packet. This means that each packet will then be sent to the controller. The controller will then do one of two different things. The first is to decide that the controller does not need to see any of the other packets that will be in that flow. In this case, the controller will provide the switch with the forwarding rule that will apply actions to all future packets that belong to this flow.

If the controller does want to process other packets in the flow, it will instruct the switch what to do with the current packet. However, no rules will be installed in the switch's flow table. This means that the next time that a packet with the same header is received by the switch, it will again be forwarded to the controller.

This selective processing of flow packets allows high-level policy rules to be expressed in low-level switch-level actions.

9.7.3 Consistent updates with Frenetic

Modern networks experience significant change over time. Links fail, network traffic increases, and maintenance must take place. When these changes happen, the network policy that is currently used by the network needs to be changed. This opens the possibility that changing the network policy could result in forwarding loops, security breaches, or transient outages. To avoid these conditions, changes must be applied to the network in a consistent manner.

Frenetic supports what researchers call a "per packet consistent update" (Foster et al., 2013). This means is that if the network starts to process a packet using network policy set 1, and while the packet is traversing the network, the network policy is updated to be policy 2, then the packet will be processed using policy 1 throughout its journey. This ensures that no loops or other conflicts will occur. Because the Frenetic system takes care of this, the network programmer can assume that he or she can create applications that are both reliable and dynamic.

The Frenetic system ensures that per packet consistency is achieved using a two-phase update system that marks packets (Foster et al., 2013). When a packet is received into the network, it is stamped with the version of the network policy that is in effect when the packet was received. As the packet travels through the rest of the network, this version number is tested at each switch that processes the packet. Each switch in the network has two copies of the network policy—the current one and the previous one. The switch has the ability to process a packet using either policy. The decision regarding which network policy to use is made based on the packet's policy stamp. Once all of the packets that were stamped with the old policy information have left the network, the controller visits each switch, deletes the old policy information, and updates the switch to use the new policy information.

The Frenetic system also supports a per flow consistent update (Foster et al., 2013). This ensures that a stream of packets that make up a flow are all processed using the same network policy. To implement this feature, the runtime system can use soft time-outs to cause the rules that were processing the old configuration to be removed and be replaced with pre-installed rules for the new configuration.

The ability to program an SDN environment is a critical feature of future networks. To realize the true potential of SDN, it must be possible to create network applications that can make use of SDN topology information and consistently update the switches in the network. The Frenetic project has created a way to accomplish this. The challenges of abstracting how the SDN can be queried and updates made in a consistent way to avoid conflicts have been solved. The foundation of tomorrow's SDN applications has already been laid.

9.8   Elastic and distributed SDN controllers

A significant advantage of SDN is that it can provide a single, unified view of the entire network. This view is provided by and maintained by a centralized controller. However, the use of a centralized controller quickly brings up the two issues of scalability and reliability: Can the controller keep up with the growth of the network? Researchers have been looking into how best to create a logically centralized, but physically distributed, controller that can solve both of these problems.

A key issue with any distributed controller is that it may be possible for any one component of the controller to become overwhelmed by network traffic. In many distributed controller designs, network SDN switches are assigned to a given controller. This static mapping creates two issues. If a given controller becomes overloaded because of a network traffic surge, then it may not be able to process all of the flows for which it is responsible. In addition, the reverse situation may arise. The controller may become underutilized and be able to support more switches. Neither of these scenarios is ideal.

To address these distributed controller issues, researchers have developed a distributed controller architecture called ElastiCon (Dixit et al., 2013). In this architecture, the load on a controller is dynamic: An overloaded controller can have its load shifted to another controller. In addition, more controllers can be brought online in the event that additional controller capacity is needed. If the load on the controller shrinks, then controllers can be removed from the system.

To create a distributed controller that can adapt to the load that is currently being applied to it, there are three operations that must be supported by the system:

1. The system will need to periodically be load balanced to achieve the optimal switch-to-controller ratio.
2. When the load grows larger than can be handled by the current pool of controllers, additional controllers will have to be added to the system. Also, once the new controller has been added to the system, switches will have to be migrated to the new controller.
3. When the load falls below a predefined level, then the pool of controllers will need to shrink. Those switches that are being managed by the controllers that are being removed will have to be migrated to controllers that remain active.

The major components that make up the ElastiCon distributed controller are shown in Figure 9.5 (Dixit et al., 2013). The architecture consists of the following components:

- There are multiple controller nodes that communicate with each other to control the network.
- Each controller node has a core control module that is responsible for executing all of the functions that a centralized controller would execute. These functions include connecting to an SDN switch and management of the communications that occur between an SDN switch and a higher-level application. This module is also responsible for working with other controllers to determine which controller is to be the MASTER controller. It will also handle the logic that is required to move switches between controllers.
- The physical network infrastructure consists of the SDN switches and the links that are used to carry network traffic. Each switch

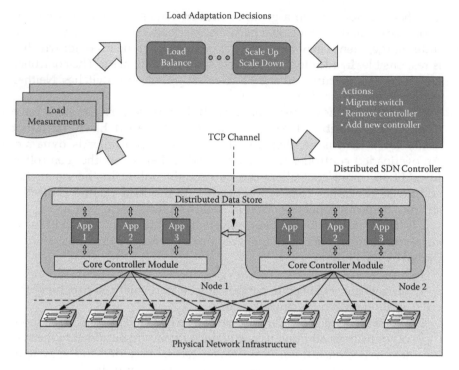

Figure 9.5 Architecture of the ElastiCon distributed controller.

can connect to one or more nodes. If a switch is connected to multiple controllers, one controller will be the MASTER and the rest will be SLAVEs.

- The distributed data store is responsible for creating a logical single centralized controller. This database stores all of the SDN switch-specific information that will be needed by the distributed controllers.

- The Transmission Control Protocol (TCP) channel is used by each controller to talk with other distributed controllers to communicate with a switch that is attached to another controller or to coordinate the transfer of a switch to another controller.

- The application modules are responsible for implementing the controller's network application logic. These applications are responsible for managing the switches that have been assigned to the controller. The current state of each application will be kept in the distributed data store to help with the switch migration process and as an aid in the event of a distributed controller failure.

9.8.1 Switch migration using the ElastiCon distributed controller

Load balancing will need to be performed periodically to ensure that none of the distributed controllers becomes overloaded. The OpenFlow protocol currently does not have any mechanism built in to perform this function.

The OpenFlow protocol supports three different modes for distributed controllers that are connected to a switch (Dixit et al., 2013). The roles are as follows:

1. EQUAL
2. SLAVE
3. MASTER

The protocol specifies that only one controller can be in the MASTER role at any time. Controllers in the MASTER and the EQUAL roles are permitted to modify the state of the switch and will receive asynchronous messages from the switch. A controller in the SLAVE role has read-only access to the switch and will not receive asynchronous messages from the switch.

In studying how to implement a switching protocol, two properties have been identified that must be guaranteed at all times during the switching process (Dixit et al., 2013):

1. There must always be a controller that is connected to the SDN switch that is currently in either the MASTER or the EQUAL mode. The controller that is in the MASTER or EQUAL mode must always be allowed to complete the processing of an asynchronous message before the controller's role is changed.
2. Only one controller will be permitted to process asynchronous messages from the SDN switch at any time.

The challenge of implementing the switch migration protocol comes from the scenario in which the SDN switch has sent an asynchronous message to an attached controller that is currently in the MASTER role. If that controller has not yet replied to the switch, then the switch cannot be migrated to another controller because this would violate the first rule. Simply changing the role of the MASTER controller to be EQUAL and then making the switch's new controller be in the MASTER role could potentially violate the second rule (Dixit et al., 2013).

Investigation of best methods to extend the OpenFlow protocol to support switch migration determined that implementation of a trigger event to facilitate the communication of the migration between two controllers would be ideal. The process of transferring a switch from one controller to another without violating either of the rules requires the following four steps (Dixit et al., 2013):

1. **The target controller has its role changed to be EQUAL:** The first step of the migration process starts with the controller that currently has the MASTER role for the switch sending a command to the controller that will be receiving the switch via the TCP channel to change its mode to EQUAL.

   The controller that will be receiving the switch then sends a command to the switch telling it to change its role to be EQUAL. Once this is completed, the switch informs the receiving controller that it is now in the EQUAL role. The receiving controller then informs the MASTER controller that the role change has been completed.

2. **Create a dummy flow:** To coordinate the migration of the SDN switch from the sending controller to the receiving controller, the sending controller will now command the switch to establish a dummy flow. Both controllers will have agreed before what the dummy flow will look like. The dummy flow will be constructed so that it will never match an incoming packet header. The switch will acknowledge to the sending controller that the dummy flow has been established. Note that because the receiving controller is in the EQUAL mode, the controller receives copies of all of the messages that the switch sends to the sending controller.

   At this point, the sending controller will send the switch what is called a "barrier message." The purpose of this message is to make sure that there is no possible way that the switch could process the delete message for the dummy flow before it processes the insert message for the dummy flow. Barrier request/reply messages are used by the controller to ensure message dependencies have been met or to receive notifications for completed operations.

   Once the sending controller has this acknowledgment, the sending controller will then command the switch to delete the dummy flow. The switch will delete the dummy flow and will then send a confirmation message to the sending controller. The receiving controller will also receive a copy of this confirmation message from the switch. At this point, the receiving controller is now in control of the switch, and the sending controller will no longer send any configuration commands to that switch.

3. **Use a barrier message to flush pending SDN switch requests:** Once the migration of the switch from the sending controller to the receiving controller has occurred, the migration process is not

considered complete. Specifically, the sending controller may still have some packets that the switch is expecting it to provide instructions on how to handle.

To ensure that this situation has been cleared up, the sending controller will transmit a barrier message that, according to the OpenFlow protocol, will cause the following events to occur: First, messages that the switch has received before the barrier message must be fully processed before the barrier message, including sending any resulting replies or errors. Next, the barrier message must then be processed and a barrier reply sent. Finally, messages after the barrier may then begin processing.

After the sending controller receives the barrier reply message from the switch, the sending controller will send an end-of-migration message to the receiving controller via the TCP channel.

4. **Change receiving controller to MASTER mode:** In the final stage of the switch migration process, the receiving controller tells the SDN switch to set its mode to be MASTER. When the switch does this, it will automatically set the mode of the sending controller to be SLAVE. At this point, all future autonomous messages from the switch will only be seen and processed by the receiving controller. The receiving controller will update the distributed data store to indicate that it is now the MASTER of the switch that has been migrated.

9.8.2 Load adaptation

An important part of the ElastiCon distributed controller model is its ability to perform load adaptation. As shown in Figure 9.5, there are three parts to load adaptation (Dixit et al., 2013):

1. Load measurement
2. Adaptation decision computation
3. Migration action

Distributed controller load statistics are continuously calculated by a load estimation module that executes as part of the controller. Load statistics consist of three controller measurements: central processing unit (CPU) usage, memory usage, and the rate of network I/O. The CPU rate is considered the best measurement of the message arrival rate (Dixit et al., 2013).

Two sets of performance thresholds are established for the ElastiCon distributed controller model. The first is a high/low loading limit for individual controllers, and the second is a high/low limit for the set of active controllers as a whole. When a given controller exceeds its high threshold but the overall load on the set of active controllers has not exceeded its threshold, then load balancing will occur. Load balancing will result in switches that are being managed by the overloaded controller being migrated to other controllers operating under their high thresholds.

If the high or low limits for the set of active controllers has been exceeded, then either additional controllers will be activated or active controllers will be deactivated. Both the load conditions and the topology of the network will be taken into account when determining which controllers will be set to MASTER mode.

When a new controller is added to the set of distributed controllers, the switches that are being managed will not know how to communicate with it. The OpenFlow protocol allows a controller to set the Internet Protocol (IP) address of other controllers within a switch. A controller that will be migrating the switch to the newly activated controller will perform the IP address update, and then the switch migration procedure will be performed.

9.9   Summary

The opportunity to participate in a revolution in the field of information technology is a rare treat. The birth of mainframes, the arrival of client-server computing, and the dawn of the Internet era are all remembered today as pivotal points in the evolution of computing. The arrival of SDN appears to be a similar point of emphasis.

SDN is not just a novel new technology. Rather, it is a considered response to the networks that have emerged over the past 50 years. Pre-SDN environments are too big and too complicated for effective cost and resource management. SDN presents a migration path from legacy networking while providing innovative network functionality and control.

Ongoing research has identified SDN as a simple, reliable approach. The fundamental philosophy of SDN is the separation of the network's control plane from its data plane. Perceived SDN advantages include better utilization of expensive network links, service engineered paths, and service appliance pooling.

Many networking ideas appear attractive in theory but do not work in practice. Google has taken the lead and transformed one of the two mission-critical backbone networks that support its business so that it now uses SDN technology. Fortified by Google's experience, SDN has moved to first position as the network technology expected to excel in the future.

The level of consideration is impressive as the OpenFlow protocol is still being refined. When Google moved forward with the project to transform its network, no commercially available OpenFlow servers were available for them to use—Google built its own. Having made this investment in both time and engineering talent, Google has publically identified things it has achieved with its new SDN environment that were not possible before (Hölzle, 2012; Vahdat et al., 2012).

If Google can do it the hard way, why cannot other corporate and government network operators do the same thing? The three things needed to build SDN environments are more widely available: a more mature OpenFlow protocol and both simple routers, based on merchant silicon that support the OpenFlow protocol, and controllers, which are now available from multiple firms and as open-source projects.

Today, SDN is investigated for what it can do for packet-switched networks. Many other types of networks, from optical transport to wireless, may also benefit from what SDN has to offer. The true power of SDN will only be realized when multilayer networks are fully virtualized and work closely together to deliver a level of service to applications that was not previously attainable.

SDN holds much promise for the future because of what it can make possible for tomorrow. For the first time, SDN allows the virtualization of the entire network. This advance provides a complete model of the network that can be used for service differentiation, traffic engineering, traffic shaping, and detailed application service delivery customization. With the ability to present data that shows a comprehensive representation of what the network is both capable of and currently doing, to prototyping novel new software applications, SDN will identify many new opportunities in service provisioning. Through the tools that SDN technology provides, including the OpenFlow protocol, applications will now be able to program the network to do what they need it to do, at the time needed.

The future of SDN is emerging. Will it remain an open standard in which freely available software will be used with simple high-speed switches to create networks that are unlike anything today? Or, will today's dominant vendors create unique solutions that customers feel more comfortable in adopting because they know that the solution will work with the equipment that they already have and that support is available if something does not work for them? The SDN path will become clear in practice over the coming decades.

The networks being built tomorrow will look nothing like the networks in service today. New operational procedures, network management techniques, and security features will be needed. Tomorrow's networks will require solving new and more complex problems, enabling a greater range of network services and control for all network users.

References

S. Azodolmolky, *Software Defined Networking with Open Flow*, Packt, Birmingham, UK, 2013.

E. Banks, SDN Showdown: Examining the Differences between VMware's NSX and Cisco's ACI, *Network World*, January 6, 2014, http://www.networkworld.com/news/2014/010614-vmware-nsx-cisco-aci-277154.html (accessed May 11, 2014).

D. Bansal, S. Bailey, T. Dietz, and A. Shaikh, OpenFlow Management and Configuration Protocol (OF-Config 1.1.1), Version 1.1.1, Open Networking Foundation, March 2013, https://www.opennetworking.org/images/stories/downloads/sdn-resources/onf-specifications/openflow-config/of-config-1-1-1.pdf

M. Caesar, D. Caldwell, N. Feamster, J. Rexford, A. Shaikh, and J. van der Merwe, Design and Implementation of a Routing Control Platform, *Proceedings of the Second Conference on Symposium on Networked Systems Design and Implementation (NSDI 2005)*, Vol. 2, pp 15–28, 2005.

M. Casado, T. Garfinkel, A. Akella, M. Freedman, D. Boneh, N. McKeown, and S. Shanker, SANE: A Protection Architecture for Enterprise Networks, *Proceedings of the 15th Conference on USENIX Security*, Vol. 15, 2006, Article 10.

V. Cert, Software Defined Network: Keynote speech ONS 2013, April 16, 2013, Open Network Summits, https://www.youtube.com/watch?v=ZrUGyUiq9TI (accessed May 11, 2014).

T. Chandra, R. Griesemer, and J. Redstone, Paxos Made Live: An Engineering Perspective, *Proceedings of the 26th ACM Symposium on Principles of Distributed Computing*, ACM, 2007, pp. 398–407.

C. Hong, S. Kandula, R. Mahajan, M. Zhang, V. Gill, M. Nanduri, R. Wattenhofer, Achieving High Utilization with Software-Driven WAN, *Proceedings of the ACM SIGCOMM 2013 Conference*, Vol. 43, No. 4, pp.15–26, 2013.

J. Dix, Google's Software-Defined/OpenFlow Backbone Drives WAN Links to 100% Utilization, *Network World*, June 7, 2012, http://www.networkworld.com/news/2012/060712-google-openflow-vahdat-259965.html (accessed May 11, 2014).

A. Dixit, F. Hao, S. Mukerkee, T. Lakshman, and R. Kompella, Towards an Elastic Distributed SDN Controller, *Proceedings of the Second ACM SIGCOMM Workshop on Hot Topics in Software Defined Networking*, ACM, pp. 7–12, 2013.

J. Duffy, Insieme FAQ: A Few Key Facts, *Network World*, November 16, 2013, http://www.networkworld.com/news/2013/110613-insieme-faq-275585. html?page=2 (accessed May 11, 2014).

S. Elby, Software Defined Networks: A Carrier Perspective, Open Network Summits, October 17, 2011, http://www.youtube.com/watch?v=xsoUexvljGk; slides available: http://www.opennetsummit.org/archives/apr12/site/talks/ elby-wed.pdf (accessed May 11, 2014).

C. Ferland, How Software Defined Networks/OpenFlow Can Transform Network Performance, *IBM System Networking*, June 29, 2012, http://www.youtube. com/watch?v=VjV5_23TUzw (accessed May 11, 2014).

FindtheBest.com, How to Find Virtualization Tool. n.d. http://virtualization.find-thebest.com/guide (accessed May 11, 2014).

N. Foster, M. Freedman, A. Guha, R. Harrison, N. Katta, C. Monsanto, J. Reich, M. Reitblatt, J. Rexford, C. Schlesinger, A. Story, and D. Walker, Languages for Software Defined Networks, *IEEE Communications*, Vol. 51, pp. 128–134, 2013.

N. Foster, R. Harrison, M. Freedman, C. Monsanto, J. Rexford, A. Story, and D. Walker, Frenetic: A Network Programming Language, *Proceedings of the 16th ACM SIGPLAN International Conference on Functional Programming (ICFP '11)*, ACM, 2011, pp. 279–291.

N. Foster, M. Freedman, R. Harrison, J. Rexford, M. Meola, and D. Walker, Frenetic: A High-Level Language for OpenFlow Networks, *Proceedings of the Workshop on Programmable Routers for Extensible Services of Tomorrow (PRESTO '10)*, Article 6, ACM, 2010.

J. Glanz, Power, Pollution and the Internet, *New York Times*, p. A1, September 23, 2012, http://www.nytimes.com/2012/09/23/technology/data-centers-waste-vast-amounts-of-energy-belying-industry-image.html?smid=pl-share (accessed May 11, 2014).

S. Gringeri, N. Bitar, and T. Xia, Extending Software Defined Network Principles to Include Optical Transport, *IEEE Communications*, Vol. 51, pp. 32–40, 2013.

N. Gude, T. Koponen, J. Pettit, B. Pfaff, M. Casado, N. McKeown, and S. Shenkar, NOX: Towards an Operating System for Networks, *ACM SIGCOMM Computer Communication Review*, Vol. 38, Issue 3, pp. 105–110, 2008.

U. Hölzle, Google Open Flow: Keynote Speech ONS 2012, Open Network Summits, April 7, 2012, http://www.youtube.com/watch?v=JMkvCBOMhno; slides available: http://www.opennetsummit.org/archives/apr12/hoelzle-tue-openflow.pdf (accessed May 11, 2014).

F. Hu, ed. *Network Innovation through OpenFlow and SDN: Principles and Design*, CRC Press, Boca Raton, FL, 2014.

S. Jain, A. Kumar, S. Mandal, J. Ong, L. Poutievski, A. Singh, S. Venkata, J. Wanderer, J. Zhou, M. Zhu, J. Zolla, U. Hölzle, S. Stuart, and A. Vahdat, B4: Experience with a Globally-Deployed Software Defined WAN, *Proceedings of the SIGCOMM '13*, ACM, pp. 3–14, 2013.

D. Levin, M. Canini, S. Schmid, and A. Feldmann, Incremental SDN Deployment in Enterprise Networks, *Proceedings of the ACM SIGCOMM 2013 Conference*, Vol. 43, No. 4, pp. 473–474, 2013.

C. Matsumoto, Is Cisco's SDN Architecture Really that Special? *SDN Central*, November 8, 2013, http://www.sdncentral.com/news/is-cisco-sdn-architecture-really-that-special/2013/11/ (accessed May 11, 2014).

D. McDysan, Software Define Networking Opportunities for Transport, *IEEE Communications*, Vol. 51, pp. 28–31, 2013.

S. McGillicuddy, Virtual Network Overlay Scalability: Legit Problem or Trolling? TechTarget.com, September 9, 2013, http://searchsdn.techtarget.com/news/2240205116/Ciscos-response-to-VMware-NSX-and-the-future-of-SDN (accessed May 11, 2014).

R. McMillan, Google Serves 25 percent of North American Internet Traffic, *Wired*, July 22, 2013, http://www.wired.com/2013/07/google-internet-traffic/ (accessed May 7, 2014).

T.P. Morgan, Big Switch Uncloaks, Fires Virty Network Wares at VMware/Nicira, theregister.co.uk, November 13, 2012, http://www.theregister.co.uk/2012/11/13/big_switch_networks_sdn/ (accessed May 11, 2014).

T.P. Morgan, Juniper Open Sources Contrail SDN Software Stack, theregister.co.uk, September 16, 2013, http://www.theregister.co.uk/2013/09/16/juniper_contrail_sdn_controller_ships/ (accessed May 11, 2014).

A.C. Murray, Juniper Launches Contrail SDN Software, Goes Open Source, *Network Computing*, September 13, 2013, http://www.networkcomputing.com/data-networking-management/juniper-launches-contrail-sdn-software-g/240161354 (accessed May 11, 2014).

T. Nadeau and K. Grey, *SDN: Software Defined Networks*, O'Reilly Media, Sebastopol, CA, 2013.

C. Neagle, Google Shares Lessons Learned as Early Software-Defined Network Adopter, *Network World*, April 11, 2012, http://www.networkworld.com/news/2012/041812-google-openflow-258406.html?hpg1=bn (accessed May 11, 2014).

OpenDaylight project, home page. n.d. http://www.opendaylight.org (accessed May 11, 2014).

OpenFlow™ Conformance Testing Program, *Open Networking Foundation*, n.d. https://www.opennetworking.org/sdn-resources/onf-specifications/openflow-conformance.

Open Networking Foundation, Software-Defined Networking: The New Norm for Networks, April 12, 2012, https://www.opennetworking.org/images/stories/downloads/sdn-resources/white-papers/wp-sdn-newnorm.pdf (accessed May 11, 2014).

Open Networking Foundation, OpenFlow Switch Specification, Version 1.4.0 (Wire Protocol 0x05), October 2013, https://www.opennetworking.org/images/stories/downloads/sdn-resources/onf-specifications/openflow/openflow-spec-v1.4.0.pdf.

SNIA, Storage Virtualization—The SNIA Technical Tutorial, n.d. https://www.snia.org/education/storage_networking_primer/stor_virt (accessed May 11, 2014).

A. Tootoonchian, S. Gorbunov, Y. Ganjali, M. Casado, and R. Sherwood, On Controller Performance in Software-Defined Networks, *Proceedings of the Second USENIX Workshop, Hot Topics in Management of Internet, Cloud, and Enterprise Networks and Services (Hot-ICE '12)*, San Jose, CA, April 2012, Article 10.

US Environmental Protection Agency, Energy Star Program, Report to Congress on Server and Data Center Energy Efficiency Public Law 109-431, US Environmental Protection Agency, Energy Star Program, Santa Clara, CA, April 2, 2007, http://www.energystar.gov/ia/partners/prod_development/downloads/EPA_Datacenter_Report_Congress_Final1.pdf (accessed May 11, 2014).

A. Vahdat, SDN Stack for Service Provider Networks, Open Network Summits, April 2012, http://www.youtube.com/watch?v=ED51Ts4o3os; slides available: http://www.opennetsummit.org/archives/apr12/vahdat-wed-sdnstack.pdf.

Viking Waters, *The History of the Mainframe Computer*, n.d., http://www.vikingwaters.com/htmlpages/MFHistory.htm.

D. Ward, SDN for Service Provider Networks: Technology, Applications, and Markets: Programmable WAN is SFW, Open Network Summits, October 18, 2011, http://www.youtube.com/watch?v=GPZ9ZcruHo8 (accessed May 11, 2014).

C. Wilson, MKM: Cisco Biggest Loser in AT&T SDN Plans, LightReading, December, 30, 2013, http://www.lightreading.com/carrier-sdn/sdn-architectures/mkm-cisco-biggest-loser-in-atandt-sdn-plans/d/d-id/707122?goback=.gde_4359316_member_5823678536455581696# (accessed May 11, 2014).

Xen Project, The Hypervisor, n.d., http://www.xenproject.org/developers/teams/hypervisor.html (accessed May 11, 2014).

S. Yeganeh, A. Tootoonchian, and Y. Ganjali, On Scalability of Software-Defined Networking, *IEEE Communications*, Vol. 51, pp. 136–141, 2013.

Index

G-Scale network, 90–104
 applications, 93
 bandwidth brokering and traffic
  engineering, 96–100
 "bandwidth demand matrix," 98–99
 command line interface, 91
 deployment results, 101–104
 fast failover, 103
 Google G-Scale network
  hardware, 91–92
 Google SDN deployment, 92–104
 Google simulation environment, 101
 G-Scale, 90, 96
 I-Scale, 90
 latency-sensitive applications, 100
 network bandwidth, allocation of, 97
 new SDN functionality, 95
 OpenFlow Agent, 91
 Open Shortest Path First
  protocols, 92
 optimized routing, 96
 path assignments, 99
 Paxos, 93
 "push on green," 102
 Quagga application, 93
 Routing Information Protocol, 93
 service-level agreement, 102
 shortest-path-first algorithm, 99
implementation challenges, 104–105
 network operations center, 104
 software functionality, 104
 traffic flows, 104
lessons learned, 105–106
 need for standardization, 105
 reduction of operating costs, 106
 redundancy, 105
 value of implementing SDN
  technology, 105
motivation for solution, 83–87
 centralized traffic-engineering
  component, 86
 convergence, 87
 deterministic behavior of network, 87
 link rebuilding process, 85
 network settling time, 84
 routers, 84
 sample network, 83
 scenario, 85
 traffic flows, 83
network testing, 87–88
 centralized traffic-engineering
  component, 87
 device configuration, 88

 lab testing, 88
 network unit testing, 88
 simulating the Google WAN, 88–89
  data centers, 88
  hardware abstraction layer, 89
  network simulation, 88
  OpenFlow interface, 88
  production code, 89
GRE/UDP, *see* Generic routing
  encapsulation/user datagram
  protocol

H

HAL, *see* Hardware abstraction layer
Hardware abstraction layer (HAL), 89
Hardware-assisted full virtualization, 15
Hub central offices, 53
Hybrid cloud APIs, 59
Hydrogen, 78
Hypervisors
 control domain, 13
 Type 2, 12
 Xen virtual machine hypervisor, 45

I

iBGP, *see* Interior BGP
Implementation, 43–49
 network operating system, 43
 SDN design, 43–47
  change of approach, 44
  commodity hardware, 44
  converging routers, 45
  edge-oriented networking, 44–47
  killer app, 43
  looping, 46
  network creation, challenge of, 45
  network functionality, 45
  network intelligence, 44
  network troubleshooting, 46
  potential impact of implementing
   SDN, 44
  separation of control and data
   planes, 44
 SDN operation, 47–49
  mesh network of routers, 47
  networking ports, configuration
   of, 47
  SDN controller, 49
  virtual local-area networks, 48
Information technology (IT), 1, 9
Interior BGP (iBGP), 91, 93

Milton Keynes UK
Ingram Content Group UK Ltd.
UKHW040054071024
449327UK00019B/564